Handbook of
PLASMA PHYSICS
Kinetics and Advanced Fluid Theory

Handbook of PLASMA PHYSICS

Kinetics and Advanced Fluid Theory

Dr. E.J. Holmyard

Oxford Book Company

Jaipur ▪ New Delhi

Published By

Oxford Book Company

Regd. Off. "Bony Residency", Gate No. 2
Opp. Tilak Public School, Vishweswariya Nagar,
Gopalpura Road, Jaipur - 302018, India
Ph: 0141 - 2761280, Telefax: 2761381
email - oxfordbookjpr@gmail.com
website - oxfordbookcompany.com

Delhi Office: 102, 1st Floor, Satyam House,
4327/3, Ansari Road, Darya Ganj,
New Delhi - 110002, India
Ph: 011-45652440

Distributed by
KEYA INTERNATIONAL PRESS LLP
E-mail: keyainternationalpressllp@gmail.com

First published in India, 2024

ISBN : 9789355241047

Typeset by Shivangi Computers, Jaipur

Printed and bound in India by Replika Press Pvt. Ltd.

Contents

Preface

A plasma is a gas in which an important fraction of the atoms is *ionized*, so that the electrons and ions are separately free. When does this ionization occur? When the temperature is hot enough. Plasmas are by far the most common phase of matter in the universe, both by mass and by volume. All the stars are made of plasma, and even the space between the stars is filled with a plasma, albeit a very sparse one. In our solar system, the planet Jupiter accounts for most of the *non*-plasma, only about 0.1% of the mass and 10^{-15}% of the volume within the orbit of Pluto. Very small grains within a gaseous plasma will also pick up a net negative charge, so that they in turn may act like a very heavy negative ion component of the plasma. In physics and chemistry, plasma is a state of matter similar to gas in which a certain portion of the particles are ionized. Heating a gas mayionize its molecules or atoms (reduce or increase the number of electrons in them), thus turning it into a plasma, which contains chargedparticles: positive ions and negative electrons or ions. Ionization can be induced by other means, such as strong electromagnetic field applied with a laser or microwave generator, and is accompanied by the dissociation of molecular bonds, if present. The presence of a non-negligible number of charge carriers makes the plasma electrically conductive so that it responds strongly toelectromagnetic fields. Plasma, therefore, has properties quite unlike those of solids, liquids, or gases and is considered a distinct state of matter. Like gas, plasma does not have a definite shape or a definite volume unless enclosed in a container; unlike gas, under the influence of a magnetic field, it may form structures such as filaments, beams and double layer. Some common plasmas are found in stars and neon signs. In the universe, plasma is the most common state of matter for ordinary matter, most of which is in the rarefied intergalactic plasma (particularlyintracluster medium) and in stars.

Plasma temperature is commonly measured in kelvins or electronvolts and is, informally, a measure of the thermal kinetic energy per particle. Very high temperatures are usually needed to sustain ionization, which is a defining feature of a plasma. The degree of plasma ionization is determined by the "electron temperature" relative to the ionization energy, (and more weakly by the density), in a relationship called the Saha equation. At low temperatures, ions and electrons tend to recombine into bound states—atoms, and the plasma will eventually become a gas. In most cases the electrons are close enough to thermal equilibrium that their temperature is relatively well-defined, even when there is a significant deviation from a Maxwellian energy distribution function, for example, due to UV radiation, energetic particles, or strong electric fields. Because of the large difference in mass, the electrons come to thermodynamic equilibrium amongst themselves much faster than they come into equilibrium with the ions or neutral atoms. For this reason, the "ion temperature" may be very different from (usually lower than) the "electron temperature". This is especially common in weakly ionized technological plasmas, where the ions are often near the ambient temperature.

Based on the relative temperatures of the electrons, ions and neutrals, plasmas are classified as "thermal" or "non-thermal". Thermal plasmas have electrons and the heavy particles at the same temperature, i.e., they are in thermal equilibrium with each other. Non-thermal plasmas on the other hand have the ions and neutrals at a much lower temperature (normally room temperature), whereas electrons are much "hotter". A plasma is sometimes referred to as being "hot" if it is nearly fully ionized, or "cold" if only a small fraction (for example 1%) of the gas molecules are ionized, but other definitions of the terms "hot plasma" and "cold plasma" are common. Even in a "cold" plasma, the electron temperature is still typically several thousand degrees Celsius.

This Handbook will help in understanding of plasmas has led to important advances in fields as diverse as computers, lighting, waste handling, space physics, switches and relays, and lasers.

—Dr. E. J. Holmyard

Chapter 1

Introduction

Plasma

Plasma is a state of matter similar to gas in which a certain portion of the particles are ionised. Heating a gas may ionise (reduce or increase the number of electrons in) its molecules or atoms, thus turning it into a plasma, which contains charged particles: positive ions and negative electrons or ions.

Ionisation can be induced by other means, such as strong electromagnetic field applied with a laser or microwave generator, and is accompanied by the dissociation of molecular bonds, if present.

The presence of a non-negligible number of charge carriers makes the plasma electrically conductive so that it responds strongly to electromagnetic fields. Plasma, therefore, has properties quite unlike those of solids, liquids, or gases and is considered a distinct state of matter.

Like gas, plasma does not have a definite shape or a definite volume unless enclosed in a container; unlike gas, under the influence of a magnetic field, it may form structures such as filaments, beams and double layer. Some common plasmas are stars and neon signs. In the universe, plasma is the most common state of matter for ordinary matter, most of which is in the rarefied intergalactic plasma (particularly intracluster medium) and in

Common Plasmas

Plasmas are by far the most common phase of matter in the universe, both by mass and by volume. All the stars are made of plasma, and even the space between the stars is filled with a plasma, albeit a very sparse one. In our solar system, the planet Jupiter accounts for most of the *non*-plasma, only about 0.1% of the mass and $10^{-5}\%$ of the volume within the orbit of Pluto. Very small grains within a gaseous plasma will also pick up a net negative charge, so that they in turn may act like a very heavy negative ion component of the plasma.

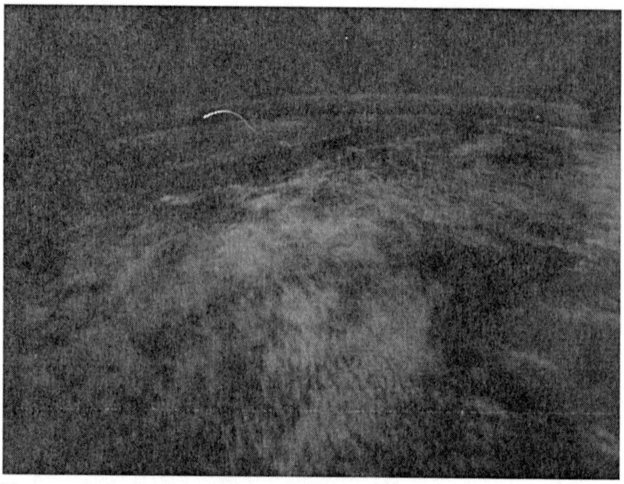

Figure 1: Plasma trail from Space Shuttle *Atlantis during* re-entry into the atmosphere, as seen from the International Space Station.

Plasma Properties and Parameters

Figure 2: Artist's rendition of the Earth's plasma fountain, showing oxygen, helium, and hydrogen ions that gush into space from regions near the Earth's poles. The faint yellow area shown above the north pole represents gas lost from Earth into space; the green area is the aurora borealis, where plasma energy pours back into the atmosphere.

Definition of a Plasma

Plasma is loosely described as an electrically neutral medium of positive and negative particles (i.e. the overall charge of a plasma is roughly zero). It is important to note that although they are unbound, these particles are not 'free'.

When the charges move they generate electrical currents with magnetic fields, and as a result, they are affected by each other's fields. This governs their collective behaviour with many degrees of freedom. A definition can have three criteria:

1. *The Plasma Approximation:* Charged particles must be close enough together that each particle influences many nearby charged particles, rather than just interacting with the closest particle (these collective effects are a distinguishing feature of a plasma). The plasma approximation is valid when the number of charge carriers within the sphere of influence (called the Debye *sphere* whose radius is the Debye screening length) of a particular particle is higher than unity to provide collective behaviour of the charged particles. The average number of particles in the Debye sphere is given by the plasma parameter, "Λ" (the Greek letterLambda).
2. *Bulk Interactions:* The Debye screening length (defined above) is short compared to the physical size of the plasma. This criterion means that interactions in the bulk of the plasma are more important than those at its edges, where boundary effects may take place. When this criterion is satisfied, the plasma is quasineutral.
3. *Plasma Frequency:* The electron plasma frequency (measuring plasma oscillations of the electrons) is large compared to the electron-neutral collision frequency (measuring frequency of collisions between electrons and neutral particles). When this condition is valid, electrostatic interactions dominate over the processes of ordinary gas kinetics.

Ranges of Plasma Parameters

Plasma parameters can take on values varying by many orders of magnitude, but the properties of plasmas with apparently disparate parameters may be very similar.

The following chart considers only conventional atomic plasmas and not exotic phenomena like quark gluon plasmas:

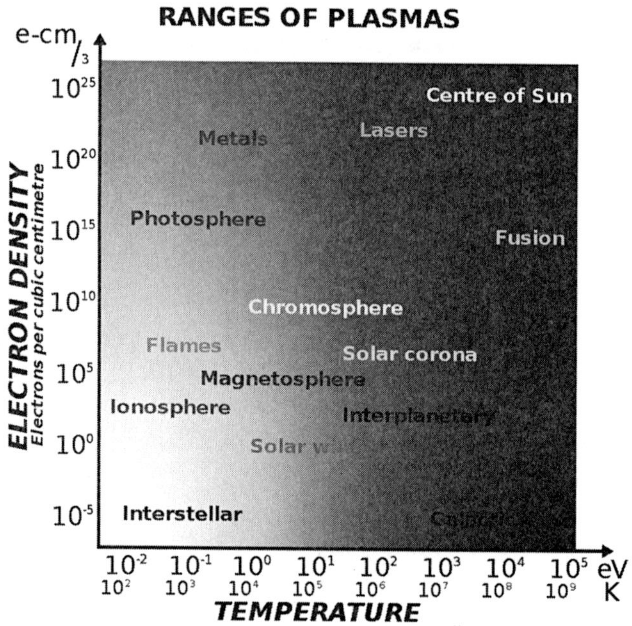

Figure 3: Range of plasmas. Density increases upwards, temperature increases towards the right. The free electrons in a metal may be considered an electron plasma.

Degree of Ionisation

For plasma to exist, ionisation is necessary. The term "plasma density" by itself usually refers to the "electron density", that is, the number of free electrons per unit volume. The degree of ionisation of a plasma is the proportion of atoms that have lost (or gained) electrons, and is controlled mostly by the temperature. Even a partially ionised gas in which as little as 1% of the particles are ionised can have the characteristics of a plasma (i.e., response to magnetic fields and high electrical conductivity). The degree of ionisation, α is defined as $\alpha = n_i/(n_i + n_a)$ where n_i is the number density of ions and n_a is the number density of neutral atoms. The *electron density* is related to this by the average charge state $\langle Z \rangle$ of the ions through $n_e = \langle Z \rangle n_i$ where n_e is the number density of electrons.

Temperatures

Plasma temperature is commonly measured in kelvins or electronvolts and is, informally, a measure of the thermal kinetic energy per particle. Very high temperatures are usually needed to sustain ionisation, which is a defining feature of a plasma. The degree of plasma ionisation is determined by the "electron temperature" relative to the ionisation energy, (and more weakly by the density), in a relationship called the Saha equation. At low temperatures, ions and electrons tend to recombine into bound states—atoms, and the plasma will eventually become a gas.

In most cases the electrons are close enough to thermal equilibrium that their temperature is relatively well-defined, even when there is a significant deviation from a Maxwellian energy distribution function, for example, due to UV radiation, energetic particles, or strong electric fields. Because of the large difference in mass, the electrons come to thermodynamic equilibrium amongst themselves much faster than they come into equilibrium with the ions or neutral atoms. For this reason, the "ion temperature" may be very different from (usually lower than) the "electron temperature". This is especially common in weakly ionised technological plasmas, where the ions are often near the ambient temperature.

Based on the relative temperatures of the electrons, ions and neutrals, plasmas are classified as "thermal" or "non-thermal". Thermal plasmas have electrons and the heavy particles at the same temperature, i.e., they are in thermal equilibrium with each other. Non-thermal plasmas on the other hand have the ions and neutrals at a much lower temperature (normally room temperature), whereas electrons are much "hotter".

A plasma is sometimes referred to as being "hot" if it is nearly fully ionised, or "cold" if only a small fraction (for example 1%) of the gas molecules are ionised, but other definitions of the terms "hot plasma" and "cold plasma" are common. Even in a "cold" plasma, the electron temperature is still typically several thousand degrees Celsius. Plasmas utilised in "plasma technology" ("technological plasmas") are usually cold in this sense.

Potentials

Since plasmas are very good conductors, electric potentials play an important role. The potential as it exists on average in the space between charged particles, independent of the question of how it can be measured, is called the "plasma potential", or the "space potential". If an electrode is inserted into a plasma, its potential will generally lie considerably below the plasma potential due to what is termed a Debye sheath. The good electrical conductivity of plasmas makes their electric fields very small. This results in the important concept of "quasineutrality", which says the density of negative charges is approximately equal to the density of positive charges over large volumes of the plasma ($n_e = \langle Z \rangle n_i$), but on the scale of the Debye length there can be charge imbalance. In the special case that *double layers* are formed, the charge separation can extend some tens of Debye lengths.

Figure 4: Lightning is an example of plasma present at Earth's surface. Typically, lightning discharges 30,000 amperes at up to 100 million volts, and emits light, radio waves, X-rays and even gamma rays. Plasma temperatures in lightning can approach ~28,000 kelvin and electron densities may exceed 10^{24} m^{-3}.

The magnitude of the potentials and electric fields must be determined by means other than simply finding the net charge density. A common example is to assume that the electrons satisfy the "Boltzmann relation":

$$n_e \propto e^{e\Phi/k_B T_e}.$$

Differentiating this relation provides a means to calculate the electric field from the density:

$$\vec{E} = (k_B T_e / e)(\nabla n_e / n_e).$$

It is possible to produce a plasma that is not quasineutral. An electron beam, for example, has only negative charges. The density of a non-neutral plasma must generally be very low, or it must be very small, otherwise it will be dissipated by the repulsive electrostatic force.

In astrophysical plasmas, Debye screening prevents electric fields from directly affecting the plasma over large distances, i.e., greater than the Debye length. However, the existence of charged particles causes the plasma to generate and can be affected by magnetic fields. This can and does cause extremely complex behaviour, such as the generation of plasma double layers, an object that separates charge over a few tens of Debye lengths. The dynamics of plasmas interacting with external and self-generated magnetic fields are studied in the academic discipline of magnetohydrodynamics.

Magnetisation

Plasma with a magnetic field strong enough to influence the motion of the charged particles is said to be magnetised. A common quantitative criterion is that a particle on average completes at least one gyration around the magnetic field before making a collision, i.e., $\omega_{ce}/\nu_{coll} > 1$, where ω_{ce} is the "electron gyrofrequency" and ν_{coll} is the "electron collision rate". It is often the case that the electrons are magnetised while the ions are not. Magnetised plasmas are *anisotropic*, meaning that their properties in the direction parallel to the magnetic field are different from those perpendicular to it. While electric fields in plasmas are usually small due to the high conductivity, the electric field associated with a plasma moving in a magnetic field is given by $E = -v \times B$ (where E is the electric field, v is the velocity, and B is the magnetic field), and is not affected by Debye shielding.

Complex Plasma Phenomena

Although the underlying equations governing plasmas are relatively simple, plasma behaviour is extraordinarily varied and subtle: the emergence of unexpected behaviour from a simple model is a typical feature of a complex system. Such systems lie in some sense on the boundary between ordered and disordered behaviour and cannot typically be described either by simple, smooth, mathematical functions, or by pure randomness. The spontaneous formation of interesting spatial features on a wide range of length scales is one manifestation of plasma complexity. The features are interesting, for example, because they are very sharp, spatially intermittent (the distance between features is much larger than the features themselves), or have a fractal form. Many of these features were first studied in the laboratory, and have subsequently been recognised throughout the universe. Examples of complexity and complex structures in plasmas include:

Filamentation

Striations or string-like structures are seen in many plasmas, like the plasma ball, the aurora, lightning, electric arcs, solar flares, and supernova remnants. They are sometimes associated with larger current densities, and the interaction with the magnetic field can form a magnetic rope structure. High power microwave breakdown at atmospheric pressure also leads to the formation of filamentary structures.

Filamentation also refers to the self-focusing of a high power laser pulse. At high powers, the nonlinear part of the index of refraction becomes important and causes a higher index of refraction in the centre of the laser beam, where the laser is brighter than at the edges, causing a feedback that focuses the laser even more. The tighter focused laser has a higher peak brightness (irradiance) that forms a plasma. The plasma has an index of refraction lower than one, and causes a defocusing of the laser beam. The interplay of the focusing index of refraction, and the defocusing plasma makes the formation of a long filament of plasma that can be micrometers to kilometres in length.

Shocks or Double Layers

Plasma properties change rapidly (within a few Debye lengths) across a two-dimensional sheet in the presence of a (moving) shock or (stationary) double layer. Double layers involve localised charge separation, which causes a large potential difference across the layer, but does not generate an electric field outside the layer. Double layers separate adjacent plasma regions with different physical characteristics, and are often found in current carrying plasmas. They accelerate both ions and electrons.

Electric Fields and Circuits

Quasineutrality of a plasma requires that plasma currents close on themselves in electric circuits. Such circuits follow Kirchhoff's circuit laws and possess a resistance and inductance. These circuits must generally be treated as a strongly coupled system, with the behaviour in each plasma region dependent on the entire circuit. It is this strong coupling between system elements, together with nonlinearity, which may lead to complex behaviour. Electrical circuits in plasmas store inductive (magnetic) energy, and should the circuit be disrupted, for example, by a plasma instability, the inductive energy will be released as plasma heating and acceleration. This is a common explanation for the heating that takes place in the solar corona. Electric currents, and in particular, magnetic-field-aligned electric currents (which are sometimes generically referred to as "Birkeland currents"), are also observed in the Earth's aurora, and in plasma filaments.

Cellular Structure

Narrow sheets with sharp gradients may separate regions with different properties such as magnetisation, density and temperature, resulting in cell-like regions. Examples include the magnetosphere, heliosphere, and heliospheric current sheet. Hannes Alfvén wrote: "From the cosmological point of view, the most important new space research discovery is probably the cellular structure of space. As has been seen in every region of space accessible to in situ measurements, there are a number of 'cell walls', sheets of electric currents, which divide space into compartments with different magnetisation, temperature, density, etc."

Critical Ionisation Velocity

The critical ionisation velocity is the relative velocity between an ionised plasma and a neutral gas, above which a runaway ionisation process takes place. The critical ionisation process is a quite general mechanism for the conversion of the kinetic energy of a rapidly streaming gas into ionisation and plasma thermal energy. Critical phenomena in general are typical of complex systems, and may lead to sharp spatial or temporal features.

Ultracold Plasma

Ultracold plasmas are created in a magneto-optical trap (MOT) by trapping and cooling neutral atoms, to temperatures of 1 mK or lower, and then using another laser to ionise the atoms by giving each of the outermost electrons just enough energy to escape the electrical attraction of its parent ion.

One advantage of ultracold plasmas are their well characterised and tunable initial conditions, including their size and electron temperature. By adjusting the wavelength of the ionising laser, the kinetic energy of the liberated electrons can be tuned as low as 0.1

K, a limit set by the frequency bandwidth of the laser pulse. The ions inherit the millikelvin temperatures of the neutral atoms, but are quickly heated through a process known as disorder induced heating (DIH). This type of non-equilibrium ultracold plasma evolves rapidly, and displays many other interesting phenomena.

One of the metastable states of a strongly nonideal plasma is Rydberg matter, which forms upon condensation of excited atoms.

Non-neutral Plasma

The strength and range of the electric force and the good conductivity of plasmas usually ensure that the densities of positive and negative charges in any sizeable region are equal ("quasineutrality"). A plasma with a significant excess of charge density, or, in the extreme case, is composed of a single species, is called a non-neutral plasma. In such a plasma, electric fields play a dominant role. Examples are charged particle beams, an electron cloud in a Penning trap and positron plasmas.

Dusty Plasma and Grain Plasma

A dusty plasma contains tiny charged particles of dust (typically found in space), which also behave like a plasma. A plasma that contains larger particles is called grain plasma.

Mathematical Descriptions

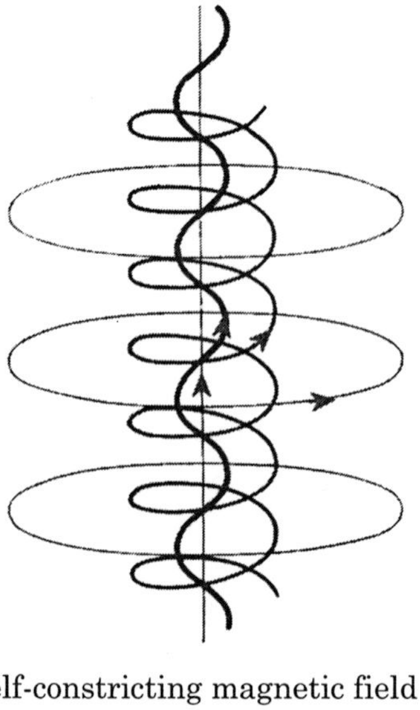

Figure 5: The complex self-constricting magnetic field lines and current paths in a field-aligned Birkeland current that can develop in a plasma.

To completely describe the state of a plasma, we would need to write down all the particle locations and velocities and describe the electromagnetic field in the plasma region. However, it is generally not practical or necessary to keep track of all the particles in a plasma. Therefore, plasma physicists commonly use less detailed descriptions, of which there are two main types:

Fluid Model

Fluid models describe plasmas in terms of smoothed quantities, like density and averaged velocity around each position. One simple fluid model, magnetohydrodynamics, treats the plasma as a single fluid governed by a combination of Maxwell's equations and the Navier–Stokes equations. A more general description is the two-fluid plasma picture, where the ions and electrons are described separately. Fluid models are often accurate when collisionality is sufficiently high to keep the plasma velocity distribution close to a Maxwell–Boltzmann distribution. Because fluid models usually describe the plasma in terms of a single flow at a certain temperature at each spatial location, they can neither capture velocity space structures like beams or double layers, nor resolve wave-particle effects.

Kinetic Model

Kinetic models describe the particle velocity distribution function at each point in the plasma and therefore do not need to assume a Maxwell–Boltzmann distribution. A kinetic description is often necessary for collisionless plasmas. There are two common approaches to kinetic description of a plasma. One is based on representing the smoothed distribution function on a grid in velocity and position. The other, known as the particle-in-cell (PIC) technique, includes kinetic information by following the trajectories of a large number of individual particles. Kinetic models are generally more computationally intensive than fluid models. The Vlasov equation may be used to describe the dynamics of a system of charged particles interacting with an electromagnetic field. In magnetised plasmas, a gyrokinetic approach can substantially reduce the computational expense of a fully kinetic simulation.

Artificial Plasmas

Most artificial plasmas are generated by the application of electric and/or magnetic fields. Plasma generated in a laboratory setting and for industrial use can be generally categorised by:

- The type of power source used to generate the plasma—DC, RF and microwave
- The pressure they operate at—vacuum pressure (< 10 mTorr or 1 Pa), moderate pressure (~ 1 Torr or 100 Pa), atmospheric pressure (760 Torr or 100 kPa)
- The degree of ionisation within the plasma—fully, partially, or weakly ionised
- The temperature relationships within the plasma—thermal plasma ($T_e = T_{ion} = T_{gas}$), non-thermal or "cold" plasma ($T_e >> T_{ion} = T_{gas}$)
- The electrode configuration used to generate the plasma
- The magnetisation of the particles within the plasma—magnetised (both ion and electrons are trapped in Larmor orbits by the magnetic field), partially magnetised (the electrons but not the ions are trapped by the magnetic field), non-magnetised (the magnetic field is too weak to trap the particles in orbits but may generate Lorentz forces)
- The application

Generation of Artificial Plasma

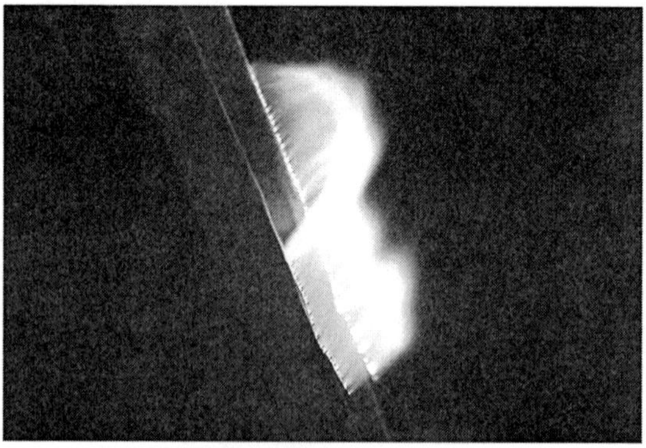

Figure 6: Artificial plasma produced in air by a Jacob's Ladder

Just like the many uses of plasma, there are several means for its generation, however, one principle is common to all of them: there must be energy input to produce and sustain it. For this case, plasma is generated when an electrical current is applied across a dielectric gas or fluid (an electrically non-conducting material) as can be seen in the image below, which shows a discharge tube as a simple example (DC used for simplicity).

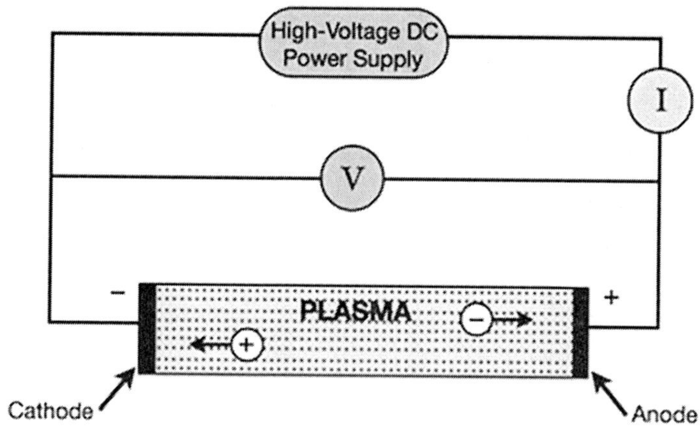

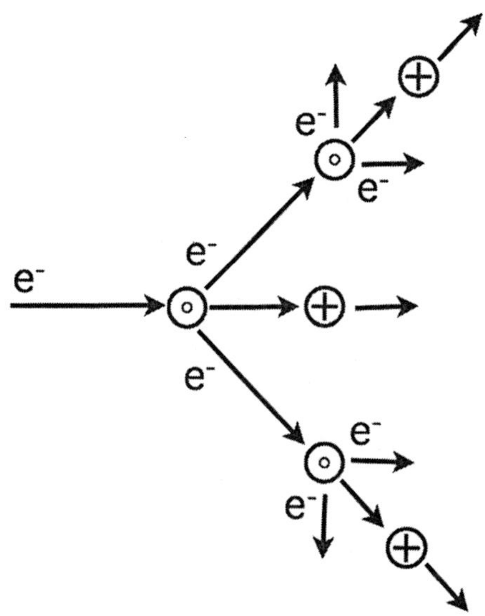

Figure 7: Cascade process of ionisation. Electrons are 'e'", neutral atoms 'o', and cations '+'.

The potential difference and subsequent electric field pull the bound electrons (negative) toward the anode (positive electrode) while the cathode (negative electrode) pulls the nucleus. As the voltage increases, the current stresses the material (by electric polarisation) beyond its dielectric limit (termed strength) into a stage of electrical breakdown, marked by an electric spark, where the material transforms from being an insulator into a conductor (as it becomes increasingly ionised). This is a stage of avalanching ionisation, where collisions between electrons and neutral gas atoms create more ions and electrons. The first impact of an electron on an atom results in one ion and two electrons. Therefore, the number of charged particles increases rapidly (in the millions) only "after about 20 successive sets of collisions", mainly due to a small mean free path (average distance travelled between collisions).

With ample current density and ionisation, this forms a luminous electric arc (essentially lightning) between the electrodes. Electrical resistance along the continuous electric arc creates heat, which ionises more gas molecules (where degree of ionisation is determined by temperature), and as per the sequence: solid-liquid-gas-plasma, the gas is gradually turned into a thermal plasma. A thermal plasma is in thermal equilibrium, which is to say that the temperature is relatively homogeneous throughout the heavy particles (i.e. atoms, molecules and ions) and electrons. This is so because when thermal plasmas are generated, electrical energy is given to electrons, which, due to their great mobility and large numbers, are able to disperse it rapidly and by elastic collision (without energy loss) to the heavy particles.

Examples of Industrial/Commercial Plasma

Because of their sizable temperature and density ranges, plasmas find applications in many fields of research, technology and industry. For example, in: industrial and extractive metallurgy, surface treatments such as thermal spraying (coating), etching in microelectronics, metal cutting and welding; as well as in everyday vehicle exhaust cleanup and fluorescent/luminescent lamps, while even playing a part in supersonic combustion engines for aerospace engineering.

Low-Pressure Discharges

- *Glow discharge plasmas*: non-thermal plasmas generated by the application of DC or low frequency RF (<100 kHz) electric field to the gap between two metal electrodes. Probably the most common plasma; this is the type of plasma generated within fluorescent light tubes.
- *Capacitively coupled plasma (CCP)*: similar to glow discharge plasmas, but generated with high frequency RF electric fields, typically 13.56 MHz. These differ from glow discharges in that the sheaths are much less intense. These are widely used in the microfabrication and integrated circuit manufacturing industries for plasma etching and plasma enhanced chemical vapour deposition.

- *Inductively coupled plasma (ICP)*: similar to a CCP and with similar applications but the electrode consists of a coil wrapped around the discharge volume that inductively excites the plasma.
- *Wave heated plasma*: similar to CCP and ICP in that it is typically RF (or microwave), but is heated by both electrostatic and electromagnetic means. Examples are helicon discharge, electron cyclotron resonance (ECR), and ion cyclotron resonance (ICR). These typically require a coaxial magnetic field for wave propagation.

Atmospheric Pressure

- *Arc discharge:* this is a high power thermal discharge of very high temperature (~10,000 K). It can be generated using various power supplies. It is commonly used in metallurgical processes. For example, it is used to melt rocks containing Al_2O_3 to produce aluminium.
- *Corona discharge:* this is a non-thermal discharge generated by the application of high voltage to sharp electrode tips. It is commonly used in ozone generators and particle precipitators.
- *Dielectric barrier discharge (DBD):* this is a non-thermal discharge generated by the application of high voltages across small gaps wherein a non-conducting coating prevents the transition of the plasma discharge into an arc. It is often mislabelled 'Corona' discharge in industry and has similar application to corona discharges. It is also widely used in the web treatment of fabrics. The application of the discharge to synthetic fabrics and plastics functionalises the surface and allows for paints, glues and similar materials to adhere.
- *Capacitive discharge:* this is a nonthermal plasma generated by the application of RF power (e.g., 13.56 MHz) to one powered electrode, with a grounded electrode held at a small separation distance on the order of 1 cm. Such discharges are commonly stabilised using a noble gas such as helium or argon.

History

Plasma was first identified in a Crookes tube, and so described by Sir William Crookes in 1879 (he called it "radiant matter"). The nature of the Crookes tube "cathode ray" matter was subsequently identified by British physicist Sir J.J. Thomson in 1897. The term "plasma" was coined by Irving Langmuir in 1928, perhaps because the glowing discharge moulds itself to the shape of the Crooks tube. Langmuir described his observations as:

Except near the electrodes, where there are *sheaths* containing very few electrons, the ionised gas contains ions and electrons in about equal numbers so that the resultant space charge is very small. We shall use the name *plasma* to describe this region containing balanced charges of ions and electrons.

Fields of Active Research

Figure 8: Hall effect thruster. The electric field in a plasma double layer is so effective at accelerating ions that electric fields are used in ion drives.

This is just a partial list of topics. A more complete and organised list can be found on the web site Plasma science and technology.

- Plasma theory
  - Plasma equilibria and stability
  - Plasma interactions with waves and beams
  - Guiding centre
  - Adiabatic invariant
  - Debye sheath
  - Coulomb collision
- Plasmas in nature
  - The Earth's ionosphere
  - Northern and southern (polar) lights
  - Space plasmas, e.g. Earth's plasmasphere (an inner portion of the magnetosphere dense with plasma)

  - Astrophysical plasma, Interplanetary medium
- Industrial plasmas
  - Plasma chemistry
  - Plasma processing
  - Plasma spray
- Plasma display

- Plasma sources
- Dusty plasmas
- Plasma diagnostics
  - Thomson scattering
  - Langmuir probe
  - Spectroscopy
  - Interferometry
  - Ionospheric heating
  - Incoherent scatter radar
- Plasma applications
  - Fusion power
- Magnetic fusion energy (MFE) —tokamak, stellarator,reversed field pinch,magnetic mirror, dense plasma focus
- Inertial fusion energy (IFE) (also Inertial confinement fusion — ICF)
- Plasma-based weaponry
  - Ion implantation
  - Ion thruster
  - MAGPIE (short for Mega *Ampere Generator for Plasma Implosion Experiments*)
  - Plasma ashing
  - Food processing (nonthermal plasma, aka "cold plasma")
  - Plasma arc waste disposal, convert waste into reusable material with plasma.
  - Plasma acceleration
  - Plasma medicine (e. g. Dentistry)
  - Plasma window.

Kinetics

Kinetics is a term classical mechanics that is concerned with the relationship between the motion of bodies and its causes, namely forces and torques. Since the mid-20th century, the term "dynamics" (or "analytical dynamics") has largely superseded "kinetics" in physics text books; the term "kinetics" is still used in engineering.

The Kinetics is deduced from Kinematics by the introduction of the concept of mass.

In plasma physics, kinetics refers to the study of continua in velocity space. This is usually in the context of non-thermal (non-Maxwellian) velocity distributions, or processes that perturb thermal distributions. These plasmas cannot be adequately described with fluid equations. Such plasmas are termed kinetic plasmas.

The term *kinetics* is also used to refer to chemical kinetics, particularly in chemical physics and physical chemistry. In such uses, a qualifier is often used or implied, for example: "physical kinetics", "crystal growth kinetics", and so on.

Fluid Mechanics

Fluid mechanics is the study of fluids and the forces on them. (Fluids include liquids, gases, and plasmas.) Fluid mechanics can be divided into fluid statics, the study of fluids at rest; fluid kinematics, the study of fluids in motion; and fluid dynamics, the study of the effect of forces on fluid motion. It is a branch of continuum mechanics, a subject which models matter without using the information that it is made out of atoms, that is, it models matter from a macroscopic viewpoint rather than from a microscopic viewpoint. Fluid mechanics, especially fluid dynamics, is an active field of research with many unsolved or partly solved problems. Fluid mechanics can be mathematically complex. Sometimes it can best be solved by numerical methods, typically using computers. A modern discipline, called computational fluid dynamics (CFD), is devoted to this approach to solving fluid mechanics problems. Also taking advantage of the highly visual nature of fluid flow is particle image velocimetry, an experimental method for visualising and analysing fluid flow.

Brief History

The study of fluid mechanics goes back at least to the days of ancient Greece, when Archimedes investigated fluid statics and buoyancy and formulated his famous law known now as the Archimedes' Principle. Rapid advancement in fluid mechanics began with Leonardo da Vinci (observation and experiment), Evangelista Torricelli (barometer), Isaac Newton (viscosity) and Blaise Pascal (hydrostatics), and was continued by Daniel Bernoulli with the introduction of mathematical fluid dynamics in *Hydrodynamica* (1738). Inviscid flow was further analysed by various mathematicians (Leonhard Euler, d'Alembert, Lagrange, Laplace, Poisson) and viscous flow was explored by a multitude of engineers including Poiseuille and Gotthilf Heinrich Ludwig Hagen. Further mathematical justification was provided by Claude-Louis Navier and George Gabriel Stokes in the Navier–Stokes equations, and boundary layers were investigated (Ludwig Prandtl, Theodore von Kármán), while various scientists (Osborne Reynolds, Andrey Kolmogorov, Geoffrey Ingram Taylor) advanced the understanding of fluid viscosity and turbulence.

Relationship to Continuum Mechanics

Fluid mechanics is a subdiscipline of continuum mechanics, as illustrated in the following table.

Continuum mechanics The study of the physics of continuous materials	**Solid mechanics** The study of the physics of continuous materials with a defined rest shape.	**Elasticity** Describes materials that retu shape after an applied stress.
		Plasticity Describes materials that permanently deform after a sufficient applied stress.
	Fluid mechanics The study of the physics of continuous materials which take the shape of their container.	Non-Newtonian fluids
		Newtonian fluids

In a mechanical view, a fluid is a substance that does not support shear stress; that is why a fluid at rest has the shape of its containing vessel. A fluid at rest has no shear stress.

Assumptions

Like any mathematical model of the real world, fluid mechanics makes some basic assumptions about the materials being studied. These assumptions are turned into equations that must be satisfied if the assumptions are to be held true. For example, consider an incompressible fluid in three dimensions. The assumption that mass is conserved means that for any fixed closed surface (such as a sphere) the rate of mass passing from *outside* to *inside* the surface must be the same as rate of mass passing the other way. (Alternatively, the mass *inside* remains constant, as does the mass *outside*). This can be turned into an integral equation over the surface.

Fluid mechanics assumes that every fluid obeys the following:

- Conservation of mass
- Conservation of energy
- Conservation of momentum
- The *continuum hypothesis*, detailed below.

Further, it is often useful (at subsonic conditions) to assume a fluid is incompressible – that is, the density of the fluid does not change.

Similarly, it can sometimes be assumed that the viscosity of the fluid is zero (the fluid is *inviscid*). Gases can often be assumed to be inviscid. If a fluid is viscous, and its flow contained in some way (e.g. in a pipe), then the flow at the boundary must have zero velocity. For a viscous fluid, if the boundary is not porous, the shear forces between the fluid and the boundary results also in a zero velocity for the fluid at the boundary. This is called the no-slip condition. For a porous media otherwise, in the frontier of the containing vessel, the slip condition is not zero velocity, and the fluid has a discontinuous velocity field between the free fluid and the fluid in the porous media (this is related to the Beavers and Joseph condition).

The Continuum Hypothesis

Fluids are composed of molecules that collide with one another and solid objects. The continuum assumption, however, considers fluids to be continuous. That is, properties such as density, pressure, temperature, and velocity are taken to be well-defined at "infinitely" small points, defining a REV (Reference Element of Volume), at the geometric order of the distance between two adjacent molecules of fluid. Properties are assumed to vary continuously from one point to another, and are averaged values in the REV. The fact that the fluid is made up of discrete molecules is ignored.

The continuum hypothesis is basically an approximation, in the same way planets are approximated by point particles when dealing with celestial mechanics, and therefore results in approximate solutions. Consequently, assumption of the continuum hypothesis can lead to results which are not of desired accuracy. That said, under the right circumstances, the continuum hypothesis produces extremely accurate results.

Those problems for which the continuum hypothesis does not allow solutions of desired accuracy are solved using statistical mechanics. To determine whether or not to use conventional fluid dynamics or statistical mechanics, the Knudsen number is evaluated for the problem. The Knudsen number is defined as the ratio of the molecular mean free path length to a certain representative physical lengthscale. This length scale could be, for example, the radius of a body in a fluid. (More simply, the Knudsen number is how many times its own diameter a particle will travel on average before hitting another particle). Problems with Knudsen numbers at or above unity are best evaluated using statistical mechanics for reliable solutions.

Navier–Stokes Equations

The Navier–Stokes equations (named after Claude-Louis Navier and George Gabriel Stokes) are the set of equations that describe the motion of fluid substances such as liquids and gases. These equations state that changes in momentum (force) of fluid particles depend only on the external pressure and internal viscous forces (similar to friction) acting on the fluid. Thus, the Navier–Stokes equations describe the balance of forces acting at any given region of the fluid.

The Navier–Stokes equations are differential equations which describe the motion of a fluid. Such equations establish relations among the rates of change of the variables of interest. For example, the Navier–Stokes equations for an ideal fluid with zero viscosity states that acceleration (the rate of change of velocity) is proportional to the derivative of internal pressure.

This means that solutions of the Navier–Stokes equations for a given physical problem must be sought with the help of calculus. In practical terms only the simplest cases can be solved exactly in this way. These cases generally involve non-turbulent, steady flow (flow does not change with time) in which the Reynolds number is small.

For more complex situations, such as global weather systems like El Niño or lift in a wing, solutions of the Navier–Stokes equations can currently only be found with the help of computers. This is a field of sciences by its own called computational fluid dynamics.

General form of the Equation

The general form of the Navier–Stokes equations for the conservation of momentum is:

$$\rho \frac{Dv}{Dt} = \nabla \cdot P + \rho f$$

where

- ρ is the fluid density,
- $\frac{D}{Dt}$ is the substantive derivative (also called the material derivative),
- v is the velocity vector,
- f is the body force vector, and
- P is a tensor that represents the surface forces applied on a fluid particle (the stress tensor).

Unless the fluid is made up of spinning degrees of freedom like vortices, P is a symmetric tensor. In general, (in three dimensions) P has the form:

$$P = \begin{pmatrix} \sigma_{xx} & \tau_{xy} & \tau_{xz} \\ \tau_{yx} & \sigma_{yy} & \tau_{yz} \\ \tau_{zx} & \tau_{zy} & \sigma_{zz} \end{pmatrix}$$

where

- σ are normal stresses,
- τ are tangential stresses (shear stresses).

The above is actually a set of three equations, one per dimension. By themselves, these aren't sufficient to produce a solution. However, adding conservation of mass and appropriate boundary conditions to the system of equations produces a solvable set of equations.

Newtonian versus Non-Newtonian Fluids

A Newtonian fluid (named after Isaac Newton) is defined to be a fluid whose shear stress is linearly proportional to the velocity gradient in the direction perpendicular to the plane of shear. This definition means regardless of the forces acting on a fluid, it *continues to flow*. For example, water is a Newtonian fluid, because it continues to display fluid properties no matter how much it is stirred or mixed. A slightly less rigorous definition is that the drag of a small object being moved slowly through the fluid is proportional to the force applied to the object. (Compare friction). Important fluids, like water as well as most gases, behave — to good approximation — as a Newtonian fluid under normal conditions on Earth.

By contrast, stirring a non-Newtonian fluid can leave a "hole" behind. This will gradually fill up over time – this behaviour is seen in materials such as pudding, oobleck, or sand (although sand isn't strictly a fluid). Alternatively, stirring a non-Newtonian fluid can cause the viscosity to decrease, so the fluid appears "thinner" (this is seen in non-drip paints). There are many types of non-Newtonian fluids, as they are defined to be something that fails to obey a particular property — for example, most fluids with long molecular chains can react in a non-Newtonian manner.

Equations for a Newtonian Fluid

The constant of proportionality between the shear stress and the velocity gradient is known as the viscosity. A simple equation to describe Newtonian fluid behaviour is

$$\tau = -\mu \frac{dv}{dy}$$

where

τ is the shear stress exerted by the fluid ("drag")

μ is the fluid viscosity – a constant of proportionality

$\frac{dv}{dy}$ is the velocity gradient perpendicular to the direction of shear.

For a Newtonian fluid, the viscosity, by definition, depends only on temperature and pressure, not on the forces acting upon it. If the fluid is incompressible and viscosity is constant across the fluid, the equation governing the shear stress (in Cartesian coordinates) is

$$\tau_{ij} = \mu\left(\frac{\partial v_i}{\partial x_j} + \frac{\partial v_j}{\partial x_i}\right)$$

where

τ_{ij} is the shear stress on the ith face of a fluid element in the jth direction

v_i is the velocity in the ith direction

x_j is the jth direction coordinate.

If a fluid does not obey this relation, it is termed a non-Newtonian fluid, of which there are several types.

Among fluids, two rough broad divisions can be made: ideal and non-ideal fluids. An ideal fluid really does not exist, but in some calculations, the assumption is justifiable. An Ideal fluid is non viscous-offers no resistance whatsoever to a shearing force.

One can group real fluids into Newtonian and non-Newtonian. Newtonian fluids agree with Newton's law of viscosity. Non-Newtonian fluids can be either plastic, bingham plastic, pseudoplastic, dilatant, thixotropic, rheopectic, viscoelatic.

Chapter 2

Common Forms of Plasmas

Astrophysical Plasma

An astrophysical plasma is a plasma (an ionised gas) the physical properties of which are studied as part of astrophysics. Much of the baryonic matter of the universe is thought to consist of plasma, a state of matter in which atoms and molecules are so hot, that they have ionised by breaking up into their constituent parts, negatively charged electrons and positively charged ions. Because the particles are charged, they are strongly influenced by electromagnetic forces, that is, by magnetic and electric fields.

All known astrophysical plasmas are influenced by magnetic fields. Since plasmas contain equal numbers of electrons and ions, they are electrically neutral overall and thus electric fields play a lesser dynamical role. Because plasmas are highly conductive, any charge imbalances are readily neutralised.

Observational Evidence

Astrophysical plasma may be studied in a variety of ways since they emit electromagnetic radiation across a wide range of the electromagnetic spectrum. Because astrophysical plasmas are generally hot, (meaning that they are fully ionised), electrons in the plasmas are continually emitting X-rays through a process called brems strahlung, when electrons nearly collide with atomic nuclei. This radiation may be detected with X-ray observatories, performed in the upper atmosphere or space, such as by the Chandra X-ray Observatory satellite. Astrophysical plasmas also emit radio waves and gamma rays.

Research and Investigation

Both plasma physicists and astrophysicists are interested in active galactic nuclei, because they are the astrophysical plasmas most directly related to the plasmas studied in the laboratory, and those studied in fusion power experiments.

They exhibit an array of complex magnetohydrodynamic behaviours, such as turbulence and instabilities. Although these phenomena can occur on scales as large as the galactic core, most physicists therorize that most phenomena on the largest scales do not involve plasma effects.

In Physical Cosmology

In the big bang cosmology the entire universe was a plasma prior to recombination. Afterwards, much of the universe reionised after the first quasars formed and emitted radiation which reionised most of the universe, which largely remains in plasma form. It is assumed by many scientists that very little baryonic matter is neutral.

In particular, the intergalactic medium, the interstellar medium, the interplanetary medium and solar winds are all mainly diffuse plasmas, and stars are made of dense plasma.

The study of astrophysical plasmas is part of the mainstream of academic astrophysics and is taken in account for in the standard cosmological model; however, current models indicate that plasma processes have little role to play in forming the very largest structures, such as voids, galaxy clusters and superclusters.

History

In 1913, Norwegian explorer and physicist Kristian Birkeland may have been the first to predict that space is filled with plasma. He wrote: "It seems to be a natural consequence of our points of view to assume that the whole of space is filled with electrons and flying electric ions of all kinds.

We have assumed that each stellar system in evolutions throws off electric corpuscles into space. It does not seem unreasonable therefore to think that the greater part of the material masses in the universe is found, not in the solar systems or nebulae, but in "empty" space."

In 1937, plasma physicist Hannes Alfvén argued that if plasma pervaded the universe, then it could generate a galactic magnetic field. During the 1940s and 50s, Alfvén developed magnetohydrodynamics (MHD) which enables plasmas to be modelled as waves in a fluid, for which Alfvén won the 1970 Nobel Prize for physics. MHD is a standard astronomical tool.

Interstellar Medium

In astronomy, the interstellar medium (orISM) is the matter that exists in the space between the star systems in a galaxy. This matter includes gas in ionic, atomic, and molecular form, dust, and cosmic rays.

It fills interstellar space and blends smoothly into the surrounding intergalactic space. The energy that occupies the same volume, in the form of electromagnetic radiation, is the interstellar radiation field.

The interstellar medium is composed of multiple phases, distinguished by whether matter is ionic, atomic, or molecular, and the temperature and density of the matter.

The thermal pressures of these phases are in rough equilibrium with one another. Magnetic fields and turbulent motions also provide pressure in the ISM, and are typically more important dynamically than the thermal pressure is.

In all phases, the interstellar medium is extremely dilute by terrestrial standards. In cool, dense regions of the ISM, matter is primarily in molecular form, and reaches number densities of 106 molecules cm^{-3}.

In hot, diffuse regions of the ISM, matter is primarily ionised, and the density may be as low as 10^{-4} ions cm^{-3}. Compare this with a number density of roughly 10^{22} cm^{-3} for liquid water. By mass, 99% of the ISM is gas in any form, and 1% is dust.

Of the gas in the ISM, 89% of atoms are hydrogen and 9% are helium, with 2% of atoms being elements heavier than hydrogen or helium, which are called "metals" in astronomical parlance. The hydrogen and helium are a result of primordial nucleo synthesis, while the heavier elements in the ISM are a result of enrichment in the process of stellar evolution.

The ISM plays a crucial role in astrophysics precisely because of its intermediate role between stellar and galactic scales. Stars form within the densest regions of the ISM, molecular clouds, and replenish the ISM with matter and energy through planetary nebulae, stellar winds, and supernovae.

This interplay between stars and the ISM helps determine the rate at which a galaxy depletes its gaseous content, and therefore its lifespan of active star formation.

Interstellar Matter

Table 1 shows a breakdown of the properties of the components of the ISM of the Milky Way.

Table 1: Components of the interstellar medium

Component	*Fractional Volume*	*Scale Height (pc)*	*Temperature (K)*	*Density (atoms/cm³)*	*State ofhydrogen*	*Primary observational techniques*
Molecular clouds	< 1%	80	10—20	10^2—10^6	molecular	Radio and infrare dmolecular emission and absorption lines
Cold Neutral Medium (CNM)	1—5%	100—300	50—100	20—50	neutral atomic	H I 21 cm lineabsorption
Warm Neutral Medium (WNM)	10—20%	300—400	6000—10000	0.2—0.5	neutral atomic	H I 21 cm line emission
Warm Ionized Medium (WIM)	20—50%	1000	8000	0.2—0.5	ionized	Hα emission and pulsar dispersion
H II regions	< 1%	70	8000	10^2—10^4	ionized	Hα emission and pulsar dispersion
Coronal gas Hot Ionized Medium (HIM)	30—70%	1000—3000	10^6—10^7	$10^{?4}$—$10^{?2}$	ionized (metals also highly ionized)	X-ray emission; absorption lines of highly ionized metals, primarily in theultraviolet

The Three-phase Model

Field, Goldsmith & Habing (1969) put forward the static two *phase* equilibrium model to explain the observed properties of the ISM. Their modelled ISM consisted of a cold dense phase (T < 300 K), consisting of clouds of neutral and molecular hydrogen, and a warm inter cloud phase (T ~ 10^4 K), consisting of rarefied neutral and ionised gas. McKee & Ostriker (1977) added a dynamic third phase that represented the very hot (T ~ 10^6 K) gas which had been shock heated by supernovae and constituted most of the volume of the ISM. These phases are the temperatures where heating and cooling can reach a stable equilibrium. Their paper formed the basis for further study over the past three decades. However, the relative proportions of the phases and their subdivisions are still not well known.

Structures

The ISM is turbulent and therefore full of structure on all spatial scales.

Stars are born deep inside large complexes of molecular clouds, typically a few parsecs in size. During their lives and deaths, stars interact physically with the ISM.

Stellar winds from young clusters of stars (often with giant or supergiant HII regions surrounding them) and shock waves created by supernovae inject enormous amounts of energy into their surroundings, which leads to hypersonic turbulence. The resultant structures – of varying sizes – can be observed, such as stellar wind bubbles and superbubbles of hot gas, seen by X-ray satellite telescopes or turbulent flows observed in radio telescope maps.

The Sun is currently travelling through the Local Interstellar Cloud, a denser region in the low-density Local Bubble.

Interaction with Interplanetary Medium

The interstellar medium begins where the interplanetary medium of the Solar System ends. The solar wind slows to subsonic velocities at the termination shock, 90—100 astronomical units from the Sun. In the region beyond the termination shock, called the heliosheath, interstellar matter interacts with the solar wind. Voyager 1, the furthest human-made object from the Earth (after 1998), crossed the termination shock December 16, 2004 and may eventually enter interstellar space, providing the first direct probe of conditions in the ISM (Stone et al. 2005).

Interstellar Extinction

The ISM is also responsible for extinction and reddening, the decreasing light intensity and shift in the dominant observable wavelengths of light from a star. These effects are caused by scattering and absorption of photons and allow the ISM to be observed with the naked eye in a dark sky. The apparent rifts that can be seen in the band of the Milky Way— a uniform disk of stars— are caused by absorption of background starlight by molecular clouds within a few thousand light years of Earth.

Far ultraviolet light is absorbed effectively by the neutral components of the ISM. For example, a typical absorption wavelength of atomic hydrogen lies at about 121.5 nanometres, the Lyman-alphatransition. Therefore, it is nearly impossible to see light emitted at that wavelength from a star farther than a few hundred light years from Earth, because most of it is absorbed during the trip to Earth by intervening neutral hydrogen.

Heating and Cooling

The ISM is usually far from thermodynamic equilibrium. Collisions establish a Maxwell-Boltzmann distribution of velocities, and the 'temperature' normally used to describe interstellar gas is the 'kinetic temperature', which describes the temperature at which the particles would have the observed Maxwell-Boltzmann velocity distribution in thermodynamic equilibrium. However, the interstellar radiation field is typically much weaker than a medium in thermodynamic equilibrium; it is most often roughly that of an A star (surface temperature of ~10,000 K) highly diluted. Therefore, bound levels within an atom or molecule in the ISM are rarely populated according to the Boltzmann formula (Spitzer 1978, § 2.4).

Depending on the temperature, density, and ionisation state of a portion of the ISM, different heating and cooling mechanisms determine the temperature of the gas.

Heating Mechanisms

Heating by Low-Energy Cosmic Rays: The first mechanism proposed for heating the ISM was heating by low-energy cosmic rays. Cosmic rays are an efficient heating source able to penetrate in the depths of molecular clouds. Cosmic rays transfer energy to gas through both ionisation and excitation and to free electrons through Coulomb interactions. Low-energy cosmic rays (a few MeV) are more important because they are far more numerous than high-energy cosmic rays.

Photoelectric Heating in Grains

The ultraviolet radiation emitted by hot stars can remove electrons from dust grains. The photonhits the dust grain, and some of its energy is used in overcoming the potential energy barrier (due to the possible positive charge of the grain) to remove the electron from the grain. The remainder of the photon's energy heats the grain and gives the ejected electron kinetic energy. Since the size distribution of dust grains is $n(r) \propto r^{-3.5}$, where r is the size of the dust particle, the grain area distribution is $r^2 n \propto r^{-1.5}$. This indicates that the smallest dust grains dominate this method of heating.

Photoionisation

When an electron is freed from an atom (typically from absorption of a UV photon) it carries kinetic energy away of the order: E_{photon} " $E_{ionisation}$. This heating mechanism dominates in HII regions, but is negligible in the diffuse ISM due to the relative lack of neutral carbon atoms.

X-ray Heating

X-rays remove electrons from atoms and ions, and those photoelectrons can provoke secondary ionisations. As the intensity is often low, this heating is only efficient in warm, less dense atomic medium (as the column density is small). For example in molecular clouds only hard x-rays can penetrate and x-ray heating can be ignored. This is assuming the region is not near an x-raysource such as a supernova remnant.

Chemical Heating

Molecular hydrogen (H_2) can be formed on the surface of dust grains when two H atoms (which can travel over the grain) meet. This process yields 4.48 eV of energy distributed over the rotational and vibrational modes, kinetic energy of the H_2 molecule, as well as heating the dust grain. This kinetic energy, as well as the energy transferred from de-excitation of the hydrogen molecule through collisions, heats the gas.

Grain-gas Heating

Collisions at high densities between gas atoms and molecules with dust grains can transfer thermal energy. This is not important in HII regions because UV radiation is more important. It is also not important in diffuse ionised medium due to the low density. In the neutral diffuse medium grains are always colder, but do not effectively cool the gas due to the low densities.

Grain heating by thermal exchange is very important in supernova remnants where densities and temperatures are very high.

Gas heating via grain-gas collisions is dominant deep in giant molecular clouds (especially at high densities). Far infrared radiation penetrates deeply due to the low optical depth. Dust grains are heated via this radiation and can transfer thermal energy during collisions with the gas. A measure of efficiency in the heating is given by the accommodation coefficient:

$$\alpha = \frac{T_2 - T}{T_d - T}$$

where T is the gas temperature, T_d the dust temperature, and T_2 the post-collision temperature of the gas atom/molecule. This coefficient was measured by (Burke & Hollenbach 1983) as $\alpha = 0.35$.

Other Heating Mechanisms

A variety of macroscopic heating mechanisms are present including:

- Gravitational collapse of a cloud
- Supernova explosions
- Stellar winds
- Expansion of H II regions
- Magnetohydrodynamic waves created by supernova remnants.

Cooling Mechanisms

Fine Structure Cooling: The process of fine structure cooling is dominant in most regions of the Interstellar Medium, except regions of hot gas and regions deep in molecular clouds. It occurs most efficiently with abundant atoms having fine structure levels close to the fundamental level such as: CII and OI in the neutral medium and OII, OIII, NII, NIII, NeII and NeIII in HII regions. Collisions will excite these atoms to higher levels, and they will eventually de-excite through photon emission, which will carry the energy out of the region.

Cooling by Permitted Lines

At higher temperatures, more levels than fine structure levels can be populated via collisions. For example, collisional excitation of the n=2 level of hydrogen will release a Lyα photon upon de-excitation. In molecular clouds, excitation of rotational lines of CO is important. Once a molecule is excited, it eventually returns to a lower energy state, emitting a photon which can leave the region, cooling the cloud.

The History of Knowledge of Interstellar Space

The nature of the interstellar medium has received the attention of astronomers and scientists over the centuries, and understanding of the ISM has developed. However, they first had to acknowledge the basic concept of "interstellar" space. The term appears to have been first used in print by Bacon (1626, § 354–5): "The Interstellar Skie.. hath .. so much Affinity with the Starre, that there is a Rotation of that, as well as of the Starre." Later, natural philosopher Robert Boyle (1674) discussed "The inter-stellar part of heaven, which several of the modern Epicureans would have to be empty."

Before modern electromagnetic theory, early physicists postulated that an invisible luminiferous aether existed as a medium to carry lightwaves. It was assumed that this aether extended into interstellar space, as Patterson (1862) wrote, "this efflux occasions a thrill, or vibratory motion, in the ether which fills the interstellar spaces."

The advent of deep photographic imaging allowed Edward Barnard to produce the first images of dark nebulae silhouetted against the background star field of the galaxy, while the first actual detection of cold diffuse matter in interstellar space was made by Johannes Hartmann in 1904 through the use of absorption line spectroscopy. In his historic study of the spectrum and orbit of Delta Orionis, Hartmann observed the light coming from this star and realised that some of this light was being absorbed before it reached the Earth. Hartmann reported that absorption from the "K" line of calcium appeared "extraordinarily weak, but almost perfectly sharp" and also reported the "quite surprising result that the calcium line at 393.4 nanometres does not share in the periodic displacements of the lines caused by the orbital motion of the spectroscopic binary star". The stationary nature of the line led Hartmann to conclude that the gas responsible for the absorption was not present in the atmosphere of Delta Orionis, but was instead located within an isolated cloud of matter residing somewhere along the line-of-sight to this star. This discovery launched the study of the Interstellar Medium.

In the series of investigations, Viktor Ambartsumian introduced for the first time the now commonly accepted notion that interstellar matter occurs in the form of clouds.

Following Hartmann's identification of interstellar calcium absorption, interstellar sodium was detected by Heger (1919) through the observation of stationary absorption from the atom's "D" lines at 589.0 and 589.6 nanometres towards Delta Orionis and Beta Scorpii.

Subsequent observations of the "H" and "K" lines of calcium by Beals (1936) revealed double and asymmetric profiles in the spectra of Epsilon and Zeta Orionis. These were the first steps in the study of the very complex interstellar sightline towards Orion. Asymmetric absorption line profiles are the result of the superposition of multiple absorption lines, each corresponding to the same atomic transition (for example the "K" line of calcium), but occurring in interstellar clouds with different radial velocities. Because each cloud has a different velocity (either towards or away from the observer/Earth) the absorption lines occurring within each cloud are either Blue-shifted or Red-shifted (respectively) from the lines' rest wavelength, through the Doppler Effect. These observations confirming that matter is not distributed homogeneously were the first evidence of multiple discrete clouds within the ISM.

The growing evidence for interstellar material led Pickering (1912) to comment that "While the interstellar absorbing medium may be simply the ether, yet the character of its selective absorption, as indicated by Kapteyn, is characteristic of a gas, and free gaseous molecules are certainly there, since they are probably constantly being expelled by the Sun and stars."

The same year Victor Hess's discovery of cosmic rays, highly energetic charged particles that rain onto the Earth from space, led others to speculate whether they also pervaded interstellar space. The following year the Norwegian explorer and physicist Kristian Birkeland wrote: "It seems to be a natural consequence of our points of view to assume that the whole of space is filled with electrons and flying electric ions of all kinds. We have assumed that each stellar system in evolutions throws off electric corpuscles into space. It does not seem unreasonable therefore to think that the greater part of the material masses in the universe is found, not in the solar [*sic*] systems or nebulae, but in 'empty' space" (Birkeland 1913).

Thorndike (1930) noted that "it could scarcely have been believed that the enormous gaps between the stars are completely void. Terrestrial aurorae are not improbably excited by charged particles from the Sun emitted by the Sun. If the millions of other stars are also ejecting ions, as is undoubtedly true, no absolute vacuum can exist within the galaxy."

Outer Space

Outer space is the void that exists beyond any celestial body, including the Earth. It is not completely empty, but consists of a hard vacuum containing a low density of particles: predominantly a plasma of hydrogen and helium, as well as electromagnetic radiation, magnetic fields, and neutrinos. In the space between galaxies, matter density can be as low as a few atoms of hydrogen per cubic meter. Theoretically, it also contains dark matter and dark energy. The baseline temperature, as set by background radiation left over from the Big Bang, is only 3 Kelvin, while in the corona of stars temperatures can reach over a million Kelvin. Plasma with an extremely low density and high temperature, such as warm-hot intergalactic medium and intracluster medium, accounts for most of the baryonic (ordinary) matter in outer space; local concentrations have evolved into stars and galaxies. Intergalactic outer space is the most voluminous component of the universe, but even galaxies and star systems consist almost entirely of empty space. As of yet, space travel has been limited to the vicinity of the Solar System; the remainder of outer space remains inaccessible to humans other than by passive observation with telescopes.

There is no firm boundary where space begins. However, the Kármán line, located at an altitude of 100 kilometres above sea level, is conventionally used for the purpose of space treaties and aerospace records keeping. The framework for international space law was established by the Outer Space Treaty, which was passed by the United Nations in 1963. This treaty precludes any claims of national sovereignty and allows all states to freely explore outer space. In 1979, the Moon Treaty made the surfaces of objects such as planets, as well as the orbital space around these bodies, the jurisdiction of the international community. Additional resolutions regarding outer space have been drafted by the United Nations, but these do not yet preclude the deployment of weapons into outer space.

Discovery

In 350 BC, Greek philosopher Aristotle suggested that nature *abhors a vacuum*, a principle that became known as the *horror vacui*. Based on this idea that a vacuum could not exist, in the West it was widely held for many centuries that space could not be empty. As late as the seventeenth century, the French philosopher René Descartes argued that the entirety of space must be filled.

In ancient China, there were various schools of thought concerning the nature of the heavens, some of which bear a resemblance to the modern understanding. In the second century CE, astronomer Zhang Heng became convinced that space must be infinite, extending well beyond the mechanism that supported the Sun and the stars. The surviving books of the Hsüan Yeh school said that the heavens were boundless, "empty and void of substance". Likewise, the "sun, moon, and the company of stars float in the empty space, moving or standing still".

The Italian scientist Galileo Galilei knew that air had weight and so was subject to gravity. In 1640, he demonstrated that an established force resisted the formation of a vacuum. However, it would remain for his pupil Evangelista Torricelli to create an apparatus that would produce a vacuum in 1643. At the time this experiment created a scientific sensation in Europe. The French mathematician Blaise Pascal reasoned that if the column of mercury was supported by air then the column ought to be shorter at higher altitude where the air pressure is lower. In 1648, his brother in law, Florin Périer, repeated the experiment on the Puy-de-Dôme mountain in central France and found that the column was shorter by three inches. This decrease in pressure was further demonstrated by carrying a half-full balloon up a mountain and watching it gradually inflate, then deflate upon descent.

In 1650, German scientist Otto von Guericke constructed the first vacuum pump; a device that would further refute the principle of *horror vacui*. He correctly noted that the atmosphere of the Earth surrounds the planet like a shell, with the density gradually declining with altitude. He concluded that there must be a vacuum between the Earth and the Moon.

In the fifteenth century, German theologian Nicolaus Cusanus speculated that the universe lacked a centre and a circumference. He believed that the universe, while not infinite, could not be held as finite as it lacked any bounds in which it could be contained. These ideas led to speculations as to the infinite dimension of space in the sixteenth century by the Italian philosopher Giordano Bruno. He extended the Copernican heliocentric cosmology to the concept of an infinite universe that is filled with a substance he called aether, which did not cause resistance to the motions of heavenly bodies. English philosopher William Gilbert arrived at a similar conclusion, arguing that the stars are visible to us only because they are surrounded by a thin aether or a void. This concept of an aether originated with ancient Greek philosophers, including Aristotle, who conceived of it as the medium through which the heavenly bodies moved.

The concept of a universe filled with a luminiferous aether remained in vogue among some scientists up until the twentieth century. This form of aether was viewed as the medium through which light could propagate. In 1887, the Michelson-Morley experiment was carried out as an attempt to detect the Earth's motion through this medium by looking for changes in the speed of light based on the direction of the planet's motion. However, the null result indicated something was wrong with the concept. Since then the idea of the luminiferous aether had essentially been abandoned, to be replaced by Albert Einstein's theory of special relativity. The latter held that the speed of light is a constant in a vacuum, regardless of the observer's motion or frame of reference.

The first professional astronomer to support the concept of an infinite universe was the Englishman Thomas Digges in 1576. However, the true scale of the universe remained unknown until the first successful measurement of the distance to a nearby star was performed in 1838 by the German astronomer Friedrich Bessel. He showed that the star 61 Cygni had a parallax of just 0.31 arc seconds (compared to the modern value of 0.2873). This corresponded to a distance of over 10 light years. The distance scale to the Andromeda galaxy was determined in 1923 by American astronomer Edwin Hubble when he measured the brightness of cepheid variables within that galaxy, a then-new technique discovered by Henrietta Leavitt. This established that the Andromeda galaxy, and by extension all galaxies, lay well outside the Milky Way.

The modern concept of outer space is based upon the Big Bang cosmology, which was first proposed in 1931 by the Belgian physicist Georges Lemaître. This theory holds that the observable universe originated from a very compact form that has since undergone continuous expansion. Matter that remained following the initial expansion has since undergone gravitational collapse to create stars, galaxies and other astronomical objects, leaving behind a deep vacuum that forms what is now called outer space. Given that light has a finite velocity, this concept also constrains the size of the directly observable universe. This leaves open the question as to whether the universe is finite or infinite.

The term *outward space* was used as early as 1842 by the English poet Lady Emmeline Stuart-Wortley in her poem "The Maiden of Moscow", although she employed it in a terrestrial context. The expression *outer space* was used as an astronomical term by Alexander von Humboldt in 1845. It was later popularized in the writings of HG

Wells in 1901. The shorter term *space* is actually older, first used to mean the region beyond Earth's sky in John Milton's *Paradise Lost* in 1667.

Environment

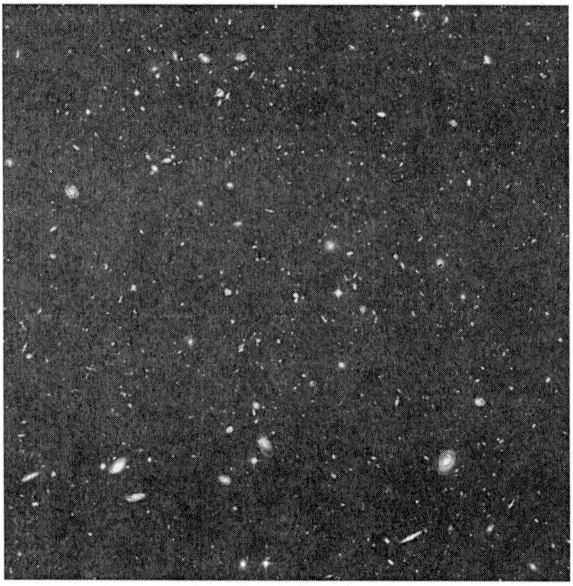

Figure 1: The Hubble Ultra-Deep Field image shows a typical section of space containing 10,000 galaxies interspaced by deep vacuum. Given the finite nature of the speed of light, this view covers the last 13 billion years of the history of outer space.

Outer space is the closest natural approximation of a perfect vacuum. It has effectively no friction, allowing stars, planets and moons to move freely along their ideal orbits. However, even the deep vacuum of intergalactic space is not devoid of matter, as it contains a few hydrogen atomsper cubic meter. By comparison, the air we breathe contains about 10 molecules per cubic meter. The sparse density of matter in outer space means that electromagnetic radiation can travel great distances without being scattered; the mean free path for a photon in intergalactic space is about 10^{23} km, or 10 billion light years. In spite of this, extinction, which is the absorption and scattering of photons by dust and gas, is an important factor in interstellar and intergalactic astronomy.

Stars, planets, asteroids and moons keep their atmospheres by gravitational attraction, and as such, atmospheres have no clearly delineated boundary: the density of atmospheric gas simply decreases with distance from the object. The Earth's atmospheric pressure drops to about 3.2×10^{-2} Pa at 100 kilometres (62 mi) of altitude, compared to 100 kPA for the IUPAC definition of standard pressure. Beyond this altitude, isotropic gas pressure rapidly becomes insignificant when compared to radiation pressure from the Sun and the dynamic pressure of the solar wind, so the definition of pressure becomes difficult to interpret. The thermosphere in this range has large gradients of pressure, temperature and composition, and varies greatly due to space weather. Astrophysicists prefer to use number density to describe these environments, employing units of particles per cubic centimetre.

On the Earth, temperature is defined in terms of the kinetic activity of the surrounding atmosphere. However, in this sense, the temperature of the vacuum can not be measured. Instead, the temperature is determined by measurement of the radiation. All of the observable Universe is filled with photons that were created during the Big Bang, which is known as the cosmic microwave background radiation (CMB). (There is quite likely a correspondingly large number of neutrinos called the cosmic neutrino background.) The current black body temperature of the background radiation is about 3 K (–270 °C; –454 °F). Some regions of outer space can contain highly energetic particles that have a much higher temperature than the CMB, such as the corona of the Sun.

Outside of a protective atmosphere and magnetic field, there are few obstacles to the passage through space of energetic subatomic particles known as cosmic rays. These particles have energies ranging from about 10^7 eV up to an extreme 10^{20} eV of ultra-high-energy cosmic rays. The peak flux of cosmic rays occurs at energies of about 10^9 eV, with approximately 87% protons, 12% helium nuclei and 1% heavier nuclei. The flux of electrons is only about 1% that of protons in all energy ranges. Cosmic rays can damage electronic components and pose a health threat to space travellers.

Effect on Human Bodies

Contrary to popular belief, a person suddenly exposed to the vacuum would not explode, freeze to death or die from boiling blood. Air would immediately leave the lungs due to the enormous pressure gradient. Any oxygen dissolved in the blood would empty into the lungs to try to equalise the partial pressure gradient. Once the deoxygenated blood arrived at the brain, death would quickly follow.

Humans and animals exposed to vacuum will lose consciousness after a few seconds and die of hypoxia within minutes. Blood and other body fluids do boil when their pressure drops below 6.3 kPa, the vapour pressure of water at body temperature. This condition is called ebullism. The steam may bloat the body to twice its normal size and slow circulation, but tissues are elastic and porous enough to prevent rupture.

Figure 2: Because of the hazards of a vacuum, astronauts must wear a pressurised spacesuit while outside their spacecraft

Ebullism is slowed by the pressure containment of blood vessels, so some blood remains liquid. Swelling and ebullism can be reduced by containment in a flight suit. Shuttle astronauts wear a fitted elastic garment called the Crew Altitude Protection Suit (CAPS) which prevents ebullism at pressures as low as 2 kPa. Water vapour would also rapidly evaporate off from exposed areas such as the lungs, cornea of the eye and mouth, cooling the body. Rapid evaporative cooling of the skin will create frost, particularly in the mouth, but this is not a significant hazard. Space may be cold, but it is mostly vacuum and transfers heat ineffectively; as a result the main temperature regulation concern for space suits is how to get rid of naturally generated body heat.

Cold or oxygen-rich atmospheres can sustain life at pressures much lower than atmospheric, as long as the density of oxygen is similar to that of standard sea-level atmosphere. The colder air temperatures found at altitudes of up to 3 kilometres (1.9 mi) generally compensate for the lower pressures there. Above this altitude, oxygen enrichment is necessary to prevent altitude sickness, and spacesuits are necessary to prevent ebullism above 19 kilometres (12 mi). Most spacesuits use only 20 kPa of pure oxygen, just enough to sustain full consciousness. This pressure is high enough to prevent ebullism, but simple evaporation of blood can still cause decompression sickness and gas embolisms if not managed.

Sudden exposure to very low pressure, such as during a rapid decompression, could cause one's lung to rupture, due to the large pressure differential between inside and outside of the chest. Even if the victim's airway is fully open, the flow of air through the windpipe may be too slow to prevent the rupture. Eardrums and sinuses may be ruptured by rapid decompression, soft tissues may bruise and seep blood, and the stress of shock will accelerate oxygen consumption leading to hypoxia. Injuries caused by rapid decompression are called barotrauma. A pressure drop as small as 13 kPa, which produces no symptoms if it is gradual, may be fatal if it occurs suddenly.

Boundary

There is no clear boundary between Earth's atmosphere and space, as the density of the atmosphere gradually decreases as the altitude increases. There are several designated scientific boundaries, namely:

- The Fédération Aéronautique Internationale has established the Kármán line at an altitude of 100 kilometres (62 mi) as a working definition for the boundary between aeronautics and astronautics. This is used because at an altitude of roughly 100 km, as Theodore von Kármán calculated, a vehicle would have to travel faster than orbital velocity in order to derive sufficient aerodynamic lift from the atmosphere to support itself.
- The United States designates people who travel above an altitude of 50 miles (80 km) as astronauts.
- NASA's mission control uses 76 miles (122 km) as their re-entry altitude (termed the Entry Interface), which roughly marks the boundary where atmospheric drag becomes noticeable (depending on the ballistic coefficient of the vehicle), thus leading shuttles to switch from steering with thrusters to maneuvering with air surfaces.

In 2009, scientists at the University of Calgary reported detailed measurements with an instrument called the Supra-Thermal Ion

Imager (an instrument that measures the direction and speed of ions), which allowed them to determine that space begins 118 kilometres (73 mi) above Earth. The boundary represents the midpoint of a gradual transition over tens of kilometres from the relatively gentle winds of the Earth's atmosphere to the more violent flows of charged particles in space, which can reach speeds well over 600 miles per hour (1,000 km/h).

Legal Status

The Outer Space Treaty provides the basic framework for international space law. This treaty covers the legal use of outer space by nation states, and includes in its definition of *outer space* the Moon and other celestial bodies.

The treaty states that outer space is free for all nation states to explore and is not subject to claims of national sovereignty. It also prohibits the deployment of nuclear weapons in outer space. The treaty was passed by the United Nations General Assembly in 1963 and signed in 1967 by the USSR, the United States of America and the United Kingdom. As of January 1, 2008 the treaty has been ratified by 98 states and signed by an additional 27 states.

Between 1958 and 2008, outer space has been the subject of multiple resolutions by the United Nations General Assembly. Of these, more than 50 have been concerning the international co-operation in the peaceful uses of outer space and preventing an arms race in space. Four additional space law treaties have been negotiated and drafted by the UN's Committee on the Peaceful Uses of Outer Space. Still, there remains no legal prohibition against deploying conventional weapons in space, and anti-satellite weapons have been successfully tested by the US, USSR and China. The 1979 Moon Treaty turned the jurisdiction of all heavenly bodies (including the orbits around such bodies) over to the international community. However, this treaty has not been ratified by any nation that currently practices manned spaceflight.

In 1976 eight equatorial states (Ecuador, Colombia, Brazil, Congo, Zaire, Uganda, Kenya, and Indonesia) met in Bogotá, Colombia. They made the "Declaration of the First Meeting of Equatorial Countries," also known as "the Bogotá Declaration", where they made a claim to control the segment of the geosyncronous orbital path corresponding to each country. These claims are not internationally accepted.

Space versus Orbit

A spacecraft enters orbit when it is travelling with sufficient horizontal velocity that the acceleration due to gravity on the spacecraft is less than or equal to the centripetal acceleration being caused by its horizontal velocity. For a low-Earth orbit, this velocity is about 7,900 m/s (28,400km/h; 17,700 mph); by contrast, the fastest aeroplane speed ever achieved (excluding speeds achieved by deorbiting spacecraft) was 2,200 m/s (7,900 km/h; 4,900 mph) in 1967 by the North American X-15.

To achieve an orbit, a spacecraft must travel faster than a sub-orbital spaceflight. The energy required to reach the velocity of low Earth orbit (32 MJ/kg) is about twenty times the energy required simply to climb to the corresponding altitude (10 kJ/(km·kg)). The minimum altitude for a stable orbit around Earth (that is, one without significant atmospheric drag) begins at around 350 kilometres (220 mi) above mean sea level. The escape velocity to pull free of Earth's gravitational field altogether and move into interplanetary space is about 11,000 m/s (39,600 km/h; 24,600 mph)

People in orbit about the Earth are able to float weightlessly because they are in free fall. However they are not outside the Earth's gravity. The orbiting spacecraft and its contents are accelerating toward Earth, but (in a perfectly circular orbit) this acceleration is a change in *direction* rather than a change in speed, bending the satellite's path from a straight line into its orbit around the earth. The weightlessness occurs because people inside a spacecraft are also orbiting the earth, at the same speed as the spacecraft that surrounds them.

Another way to describe the same situation is to consider a rotating frame of reference matching the satellite's orbit. In such a reference frame, there is a (fictitious) centrifugal force that exactly cancels the force of gravity, leaving no net force acting on the orbiting passengers. Earth's gravity reaches out far past the Van Allen belt and keeps the Moon in orbit at an average distance of 384,403 kilometres (238,857 mi).

Regions

Space is a partial vacuum: its different regions are defined by the various atmospheres and "winds" that dominate within them, and extend to the point at which those winds give way to those beyond.

Geospace extends from Earth's atmosphere to the outer reaches of Earth's magnetic field, whereupon it gives way to the solar wind of interplanetary space. Interplanetary space extends to the heliopause, whereupon the solar wind gives way to the winds of the interstellar medium. Interstellar space then continues to the edges of the galaxy, where it fades into the intergalactic void.

Geospace

Geospace is the region of outer space near the Earth. Geospace includes the upper region of the atmosphere, as well as the magnetosphere. The outer boundary of geospace is the magnetopause, which forms an interface between the planet's magnetosphere and the solar wind. The inner boundary is the ionosphere. Alternatively, geospace is the region of space between the Earth's upper atmosphere and the outermost reaches of the Earth's magnetic field. As the physical properties and behaviour of near Earth space is affected by the behaviour of the Sun and space weather, the field of geospaceis interlinked with heliophysics; the study of the Sun and its impact on the Solar System planets.

The volume of geospace defined by the magnetopause is compacted in the direction of the Sun by the pressure of the solar wind, giving it a typical subsolar distance of 10 Earth radii from the centre of the planet. However, the tail can extend outward to more than 100–200 Earth radii. The Van Allen radiation belts lie within the geospace. The region between Earth's atmosphere and the Moon is sometimes referred to as cis-lunar space. The Moon passes through geospace roughly four days each month, during which time the surface is shielded from the solar wind.

Geospace is populated by electrically charged particles at very low densities, the motions of which are controlled by the Earth's magnetic field. These plasmas form a medium from which storm-like disturbances powered by the solar wind can drive electrical currents into the Earth's upper atmosphere. During geomagnetic storms two regions of geospace, the radiation belts and the ionosphere, can become strongly disturbed. These storms increase fluxes of energetic electrons that can permanently damage satellite electronics, disrupting telecommunications and GPS technologies, and can also be a hazard to astronauts, even in low-Earth orbit. They also create aurorae seen near the magnetic poles.

Although it meets the definition of outer space, the atmospheric density within the first few hundred kilometres above the Kármán line is still sufficient to produce significant drag on satellites. Most artificial satellites operate in this region called low earth orbit and must fire their engines every few days to maintain orbit. This region contains material left over from previous manned and unmanned launches that are a potential hazard to spacecraft. Some of this debris re-enters Earth's atmosphere periodically. The drag here is low enough that it could theoretically be overcome by radiation pressure on solar sails, a proposed propulsion system for interplanetary travel.

Interplanetary

Interplanetary space, the space around the Sun and planets of the Solar System, is the region dominated by the interplanetary medium, which extends out to the heliopause where the influence of the galactic environment starts to dominate over the magnetic field and particle flux from the Sun. Interplanetary space is defined by the solar wind, a continuous stream of charged particles emanating from the Sun that creates a very tenuous atmosphere (the heliosphere) for billions of miles into space. This wind has a particle density of 5–10 protons/cm^3 and is moving at a velocity of 350–400 km/s. The distance and strength of the heliopause varies depending on the activity level of the solar wind. The discovery since 1995 of extrasolar planets means that other stars must possess their own interplanetary media.

The volume of interplanetary space is a nearly total vacuum, with a mean free path of about one astronomical unit at the orbital distance of the Earth. However, this space is not completely empty, and is sparsely filled with cosmic rays, which include ionised atomic nuclei and various subatomic particles. There is also gas, plasma and dust, small meteors, and several dozen types of organic molecules discovered to date by microwave spectroscopy.

Interplanetary space contains the magnetic field generated by the Sun. There are also magnetospheres generated by planets such as Jupiter, Saturn, Mercury and the Earth that have their own magnetic fields. These are shaped by the influence of the solar wind into the approximation of a teardrop shape, with the long tail extending outward behind the planet. These magnetic fields can trap particles from the solar wind and other sources, creating belts of magnetic particles such as the Van Allen Belts. Planets without magnetic fields, such as Mars, but excluding Venus, have their atmospheres gradually eroded by the solar wind.

Interstellar

Interstellar space is the physical space within a galaxy not occupied by stars or their planetary systems. The interstellar medium resides—by definition—in interstellar space. The average density of matter in this region is about 10^6 particles per cm^3, but this varies from a low of about 10^4–10^5 in regions of sparse matter up to about 10^8–10^{10} in dark nebula. Regions of star formation may reach 10^{12}–10^{14} particles per cm^3. Nearly 70% of this mass consists of lone hydrogen atoms. This is enriched with helium atoms as well as trace amounts of heavier atoms formed through stellar nucleo synthesis. A number of molecules can also form in interstellar space, as can tiny, 0.1 μm dust particles.

Intergalactic

Intergalactic space is the physical space between galaxies. The huge spaces between galaxy clusters are called the voids. Present estimates put the average mass density of the Universe at 5.9 protons per cubic meter, of which normal, baryonic matter has a density of one proton per four cubic meters. The density of the universe, however, is clearly not uniform; it ranges from relatively high density in galaxies (including very high density in structures within galaxies, such as planets, stars, and black holes) to conditions in vast voids that have much lower density than the universe's average.

Surrounding and stretching between galaxies, there is a rarefied plasma that is thought to possess a cosmic filamentary structure and that is slightly denser than the average density in the universe. This material is called the intergalactic medium and is mostly ionised hydrogen; i.e. a plasma consisting of equal numbers of electrons and protons. The IGM is thought to exist at a density of 10 to 100 times the average density of the universe (10 to 100 hydrogen atoms per cubic meter). In rich galaxy clusters, it reaches densities as high as 1000 times the average density of the universe (the intracluster medium).

The reason the IGM is thought to be mostly ionised gas is that its temperature is estimated to be quite high by terrestrial standards (though some parts of it are only "warm" by astrophysical standards). As gas falls into the intergalactic medium from the voids, it heats up to temperatures of 10^5 K to 10^7 K, which is high enough for the bound electrons to escape from the hydrogen nuclei upon collisions. At these temperatures, it is called the warm-hot intergalactic medium. Computer simulations indicate that on the order of half the atomic matter in the universe might exist in this warm-hot, rarefied state. When gas falls from the filamentary structures of the WHIM into the galaxy clusters at the intersections of the cosmic filaments, it can heat up even more, reaching temperatures of 10^8 K and above.

Exploration and Applications

For the majority of human history, space was explored by remote observation; initially with the unaided eye and then with the telescope. Prior to the advent of reliable rocket technology, the closest that humans had come to reaching outer space was through the use of balloon flights. In 1935, the U.S. Explorer *II* manned balloon flight had reached an altitude of 22 km (14 mi). This was greatly exceeded in 1942 when the third launch of the German A-4 rocket climbed to an altitude of about 80 km (50 mi). In 1957, the unmanned satellite *Sputnik 1* was launch by a Russian R-7 rocket, achieving Earth orbit at an altitude of 215–939 km (134–583 mi). This was followed by the first human spaceflight in 1961, when Yuri Gagarin was sent into orbit on Vostok 1. The first humans to escape Earth orbit were Frank Borman, Jim Lovell and William Anders in 1968 on board Apollo 8, which achieved lunar orbit and reached a maximum distance of 377,349 km (234,474 mi) from the Earth.

In order to explore the other planets, a spacecraft must first reach escape velocity, which will allow it to travel beyond Earth orbit. The first spacecraft to accomplish this feat was the Soviet Luna 1, which performed a fly-by of the Moon in 1959. In 1961, Venera 1 became the first planetary probe. It revealed the presence of the solar wind and performed the first fly-by of the planet Venus, although contact was lost before reaching Venus. The first successful planetary mission was the Mariner 2 fly-by of Venus in 1962. The first spacecraft to perform a fly-by of Mars was Mariner 4, which reached the planet in 1964. Since that time, unmanned spacecraft have successfully examined each of the Solar System's planets, as well their moons and many minor planets and comets. They remain a fundamental tool for the exploration of outer space.

The absence of air makes outer space (and the surface of the Moon) ideal locations for astronomy at all wavelengths of the electromagnetic spectrum, as evidenced by the spectacular pictures sent back by the Hubble Space Telescope, allowing light from about 13.7 billion years ago—almost to the time of the Big Bang—to be observed.

The deep vacuum of space could make it an attractive environment for certain industrial processes, such as those that require ultraclean surfaces.

Dusty Plasma

A dusty plasma is a plasma containing nanometre or micrometer-sized particles suspended in it. Dust particles may be charged and the plasma and particles behave as a plasma, following electromagnetic laws for particles up to about 10 nm (or 100 nm if large charges are present). Dust particles may form larger particles resulting in "grain plasmas".

Dusty plasmas are encountered in:

- Industrial processing plasmas
- Space plasmas.

Dusty plasmas are interesting because the presence of particles significantly alters the charged particle equilibrium leading to different phenomena. It is a field of current research. Electrostatic coupling between the grains can vary over a wide range so that the states of the dusty plasma can change from weakly coupled (gaseous) to crystalline. Such plasmas are of interest as a non-Hamiltonian system of interacting particles and as a means to study generic fundamental physics of self-organisation, pattern formation, phase transitions, and scaling.

Characteristics

The temperature of dust in a plasma may be quite different from its environment.

For example:

Dust Plasma Component	*Temperature*
Dust temperature	10 K
Molecular temperature	100 K
Ion temperature	1,000 K
Electron temperature	10,000 K

The electric potential of dust particles is typically 1–10 V (positive or negative). The potential is usually negative because the electrons are more mobile than the ions. The physics is essentially that of a Langmuir probe that draws no net current, including formation of a Debye sheath with a thickness of a few times the Debye length. If the electrons charging the dust grains are relativistic, then the dust may charge to several kilovolts. Field electron emission, which tends to reduce the negative potential, can be important due to the small size of the particles. The photoelectric effect and the impact of positive ions may actually result in a positive potential of the dust particles.

Dynamics

The motion of solid particles in a plasma follows the *momentum equation* for ions and electrons:

$$m\frac{dv}{dt} = mg + q(E + v \times B) - mv_c v + f$$

where m, q are the mass and charge of the particle, g is the gravitation acceleration, $m\upsilon_c v$ is due to viscosity, and f represents all other forces including radiation pressure. q (E + v x B) is the Lorentz force, where E is the electric field, v is the velocity and B is the magnetic field.

Then depending in the size of the particle, there are four categories:

1. Very small particles, where q (E + v × B) dominates over mg.
2. Small grains, where q/m ≈ -G, and plasma still plays a major role in the dynamics.
3. Large grains, where the electromagnetic term is negligible, and the particles are referred to as grains. Their motion is determined by gravity and viscosity, and the equation of motion becomes $m\upsilon_c v = mg$.
4. Large solid bodies. In centimetre and meter-sized bodies, viscosity may cause significant perturbations that can change an orbit. In kilometre-sized (or more) bodies, gravity and inertia dominate the motion.

Chapter 3

Plasma Display

A plasma display panel (PDP) is a type of flat panel display common to large TV displays 30 inches (76 cm) or larger. They are called "plasma" displays because the technology utilises small cells containing electrically charged ionised gases, or what are in essence chambers more commonly known as fluorescent lamps.

General Characteristics

Plasma displays are bright (1,000 lux or higher for the module), have a wide colour gamut, and can be produced in fairly large sizes—up to 150 inches (3.8 m) diagonally. They have a very low-luminance "dark-room" black level compared to the lighter grey of the unilluminated parts of an LCD screen (i.e. the blacks are blacker on plasmas and greyer on LCDs). LED-backlit LCD televisions have been developed to reduce this distinction.

The display panel itself is about 6 cm (2.5 inches) thick, generally allowing the device's total thickness (including electronics) to be less than 10 cm (4 inches). Plasma displays use as much power per square meter as a CRT or an AMLCD television. Power consumption varies greatly with picture content, with bright scenes drawing significantly more power than darker ones – this is also true of CRTs. Typical power consumption is 400 watts for a 50-inch (127 cm) screen. 200 to 310 watts for a 50-inch (127 cm) display when set to cinema mode. Most screens are set to 'shop' mode by default, which draws at least twice the power (around 500–700 watts) of a 'home' setting of less extreme brightness. Panasonic has greatly reduced power consumption (–1/3 of 2007 models") Panasonic states that PDPs will consume only half the power of their previous series of plasma sets to achieve the same overall brightness for a given display size. The lifetime of the latest generation of plasma displays is estimated at 100,000 hours of actual display time, or 27 years at 10 hours per day. This is the estimated time over which maximum picture brightness degrades to half the original value.

Plasma display screens are made from glass, which reflects more light than the material used to make an LCD screen. This causes glare from reflected objects in the viewing area. Companies such as Panasonic coat their newer plasma screens with an anti-glare filter material. Currently, plasma panels cannot be economically manufactured in screen sizes smaller than 32 inches. Although a few companies have been able to make plasma Enhanced-definition televisions (EDTV) this small, even fewer have made 32 in plasma HDTVs. With the trend toward large-screen television technology, the 32in screen size is rapidly disappearing. Though considered bulky and thick compared to their LCD counterparts, some sets such as Panasonic's Z1 and Samsung's B860 series are as slim as one inch thick making them comparable to LCDs in this respect.

Competing display technologies include Cathode ray tube (CRT), Organic light-emitting diode (OLED),AMLCD, Digital Light Processing DLP, SED-tv, LED display, field emission display (FED), and Quantum dot display (QLED).

Plasma Display Advantages and Disadvantages

Advantages

- Picture quality
  - o Capable of producing deeper blacks allowing for superior contrast ratio
  - o Wider viewing angles than those of LCD; images do not suffer from degradation at high angles like LCDs
  - o Less visible motion blur, thanks in large part to very high refresh rates and a faster response time, contributing to superior performance when displaying content with significant amounts of rapid motion (though newer LCD screens have similar refresh rates, but that also introduces the soap opera effect).

Disadvantages

- Picture quality
  - o Earlier generation displays were more susceptible to screen burn-in and image retention, recent models have a pixel or biter that moves the entire picture faster than is

noticeable to the human eye, which reduces the effect of burn-in but does not prevent it.

- o Earlier generation displays (circa 2006 and prior) had phosphors that lost luminosity over time, resulting in gradual decline of absolute image brightness (newer models are less susceptible to this, having lifespans exceeding 100,000 hours, far longer than older CRT technology)
- o Screen-door effects are noticeable on screen sizes smaller than 50 in, effect is more visible at shorter viewing distances.

- Other
  - o Use more electricity, on average, than an LCD TV.
  - o Do not work as well at high altitudes due to pressure differential between the gases inside the screen and the air pressure at altitude. It may cause a buzzing noise. Manufacturers rate their screens to indicate the altitude parameters.
  - o For those who wish to listen to AM radio, or are Amateur Radio operators (Hams) or Shortwave Listeners (SWL), the Radio Frequency Interference (RFI) from these devices can be irritating or disabling.
  - o Due to the strong infrared emissions inherent with the technology, standard IR repeater systems can not be used in the viewing room. A more expensive "plasma compatible" sensor must be used.

Native Plasma Television Resolutions

Fixed-pixel displays such as plasma TVs scale the video image of each incoming signal to the native resolution of the display panel. The most common native resolutions for plasma display panels are 853×480 (EDTV), 1,366×768 or 1,920×1,080 (HDTV). As a result picture quality varies depending on the performance of the video scaling processor and the upscaling and downscaling algorithms used by each display manufacturer.

Enhanced-definition Plasma Television

Early plasma televisions were enhanced-definition (ED) with a native resolution of 840×480 (discontinued) or 853×480, and downscaled their incoming High-definition video signals to match their native display resolution.

ED Resolutions

Following ED resolutions were common prior to the introduction of HD displays, but have long been phased out in favour of HD displays.

- 840×480
- 853×480

High-definition Plasma Television

Early high-definition (HD) plasma displays had a resolution of 1024x1024 and were alternate lighting of surfaces (ALiS) panels made by Fujitsu/Hitachi. These were interlaced displays, with non-square pixels.

Modern HDTV plasma televisions usually have a resolution of 1,024×768 found on many 42 inch plasma screens, 1,280×768, 1,366×768 found on 50 in, 60 in, and 65 in plasma screens, or 1,920×1,080 found in plasma screen sizes from 42 inch to 103 inch. These displays are usually progressive displays, with square pixels, and will up-scale their incoming standard-definition signals to match their native display resolution.

HD Resolutions

- 1024×1024 (discontinued)
- 1024×768
- 1280×768
- 1366×768
- 1280×1080
- 1920×1080

How Plasma Displays Work

A panel typically has millions of tiny cells in compartmentalised space between two panels of glass. These compartments, or "bulbs" or "cells", hold a mixture of noble gases and a minuscule amount of mercury. Just as in the fluorescent lamps over an office desk, when the mercury is vaporised and a voltage is applied across the cell, the gas in the cells form a plasma. With flow of electricity (electrons), some of the electrons strike mercury particles as the electrons move through the plasma, momentarily increasing the energy level of the molecule until the excess energy is shed. Mercury sheds the energy as ultraviolet (UV) photons. The UV photons then strike phosphor that is painted on the inside of the cell. When the UV photon strikes a phosphor molecule, it momentarily raises the energy level of an outer orbit electron in the phosphor molecule, moving the electron from a stable to an unstable state; the electron then sheds the excess energy as a photon at a lower energy level than UV light; the lower energy photons are mostly in the infrared range but about 40% are in the visible light range. Thus the input energy is shed as mostly heat (infrared) but also as visible light. Depending on the phosphors used, different colours of visible light can be achieved. Each pixel in a plasma display is made up of three cells comprising the primary colours of visible light. Varying the voltage of the signals to the cells thus allows different perceived colours.

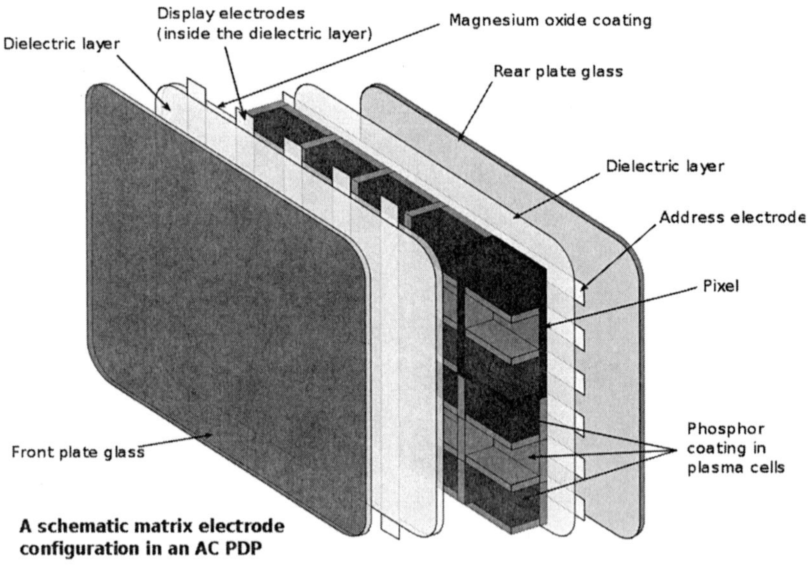

Figure 1: Composition of plasma display panel

A plasma display panel is an array of hundreds of thousands of small, luminous cells positioned between two plates of glass. Each cell is essentially a tiny neon lamp filled with rarefied neon, xenon, and other inert gases; the cells are luminous when they are electrified through "electrodes".

The long electrodes are stripes of electrically conducting material that also lie between the glass plates, in front of and behind the cells. The "address electrodes" sit behind the cells, along the rear glass plate, and can be opaque. The transparent display electrodes are mounted in front of the cell, along the front glass plate. As can be seen in the illustration, the electrodes are covered by an insulating protective layer. Control circuitry charges the electrodes that cross paths at a cell, creating a voltage difference between front and back. Some of the atoms in the gas of a cell then lose electrons and become ionised, which creates an electrically conducting plasma of atoms, free electrons, and ions. The collisions of the flowing electrons in the plasma with the inert gas atoms leads to light emission; such light-emitting plasmas are known as glow discharges.

In a monochrome plasma panel, the gas is usually mostly neon, and the colour is the characteristic orange of a neon-filled lamp (or sign). Once a glow discharge has been initiated in a cell, it can be maintained by applying a low-level voltage between all the horizontal and vertical electrodes–even after the ionising voltage is removed. To erase a cell all voltage is removed from a pair of electrodes. This type of panel has inherent memory. A small amount of nitrogen is added to the neon to increase hysteresis.

In colour panels, the back of each cell is coated with a phosphor. The ultraviolet photons emitted by the plasma excite these phosphors, which give off visible light with colours determined by the phosphor materials. This aspect is comparable to fluorescent lamps and to the neon signs that use coloured phosphors.

Every pixel is made up of three separate subpixel cells, each with different coloured phosphors. One subpixel has a red light phosphor, one subpixel has a green light phosphor and one subpixel has a blue light phosphor. These colours blend together to create the overall colour of the pixel, the same as atriad of a shadow mask CRT or colour LCD. Plasma panels use pulse-width modulation (PWM) to control brightness: by varying the pulses of current flowing through the different cells thousands of times per second, the control system can increase or decrease the intensity of each subpixel colour to create billions of different combinations of red, green and blue. In this way, the control system can produce most of the visible colours. Plasma displays use the same phosphors as CRTs, which accounts for the extremely accurate colour reproduction when viewing television or computer video images (which use an RGB colour system designed for CRT display technology).

Plasma displays should not be confused with liquid crystal displays (LCDs), another lightweight flat-screen display using very different technology. LCDs may use one or two large fluorescent lamps as a backlight source, but the different colours are controlled by LCD units, which in effect behave as gates that allow or block the passage of light from the backlight to red, green, or blue paint on the front of the LCD panel.

Contrast Ratio

Contrast ratio is the difference between the brightest and darkest parts of an image, measured in discrete steps, at any given moment. Generally, the higher the contrast ratio, the more realistic the image is (though the "realism" of an image depends on many factors including colour accuracy, luminance linearity, and spatial linearity.) Contrast ratios for plasma displays are often advertised as high as 5,000,000:1. On the surface, this is a significant advantage of plasma over most other current display technologies, a notable exception being organic light-emitting diode. Although there are no industry-wide guidelines for reporting contrast ratio, most manufacturers follow either the ANSI standard or perform a full-on-full-off test. The ANSI standard uses a checkered test pattern whereby the darkest blacks and the lightest whites are simultaneously measured, yielding the most accurate "real-world" ratings. In contrast, a full-on-full-off test measures the ratio using a pure black screen and a pure white screen, which gives higher values but does not represent a typical viewing scenario. Some displays, using many different technologies, have some "leakage" of light, through either optical or electronic means, from lit pixels to adjacent pixels so that dark pixels that are near bright ones appear less dark than they do during a full-off display. Manufacturers can further artificially improve the reported contrast ratio by increasing the contrast and brightness settings to achieve the highest test values. However, a contrast ratio generated by this method is misleading, as content would be essentially unwatchable at such settings.

Plasma is often cited as having better (i.e. darker) black levels (and higher contrast ratios), although both plasma and LCD each have their own technological challenges.

Each cell on a plasma display has to be precharged before it is due to be illuminated (otherwise the cell would not respond quickly enough) and this precharging means the cells cannot achieve a true black, whereas an LED backlit LCD panel can actually turn off parts of the screen. Some manufacturers have worked hard to reduce the precharge and the associated background glow, to the point where black levels on modern plasmas are starting to rival CRT. With LCD technology, black pixels are generated by a light polarisation method;

many panels are unable to completely block the underlying backlight. However, more recent LCD panels (particularly those using white LED illumination) can compensate by automatically reducing the backlighting on darker scenes, though this method — analogous to the strategy of noise reduction on analog audio tape — obviously cannot be used in high-contrast scenes, leaving some light showing from black parts of an image with bright parts, such as (at the extreme) a solid black screen with one fine intense bright line. This is called a "halo" effect which has been almost completely minimised on newer LED backlit LCD's with local dimming. Edgelit models cannot compete with this as the light is reflected via a light funnell to distribute the light behind the panel.

Screen Burn-in

Image burn-in occurs on CRTs and plasma panels when the same picture is displayed for long periods of time. This causes the phosphors to overheat, losing some of their luminosity and producing a "shadow" image that is visible with the power off. Burn-in is especially a problem on plasma panels because they run hotter than CRTs. Early plasma televisions were plagued by burn-in, making it impossible to use video games or anything else that displayed static images.

Plasma displays also exhibit another image retention issue which is sometimes confused with screen burn-in damage. In this mode, when a group of pixels are run at high brightness (when displaying white, for example) for an extended period of time, a charge build-up in the pixel structure occurs and a ghost image can be seen. However, unlike burn-in, this charge build-up is transient and self corrects after the image condition that caused the effect has been removed and a long enough period of time has passed (with the display either off or on).

Plasma manufacturers have tried various ways of reducing burn-in such as using gray pillarboxes, pixel orbiters and image washing routines, but none to date have eliminated the problem and all plasma manufacturers continue to exclude burn-in from their warranties.

Environmental Impact

Plasma screens have been lagging behind CRT and LCD screens in terms of energy consumption efficiency. To reduce the energy consumption, new technologies are also being found. Although it can be expected that plasma screens will continue to become more energy efficient in the future, a growing problem is that people tend to keep their old TVs running and an increasing trend to escalating screen sizes.

History

In 1936 Kálmán Tihanyi described the principle of "plasma television" and conceived the first flat-panel display system.

The monochrome plasma video display was co-invented in 1964 at the University of Illinois at Urbana-Champaign by Donald Bitzer, H. Gene Slottow, and graduate student Robert Willson for the PLATO Computer System. The original neon orange monochrome Digivue display panels built by glass producer Owens-Illinois were very popular in the early 1970s because they were rugged and needed neither memory nor circuitry to refresh the images. A long period of sales decline occurred in the late 1970s because semiconductor memory made CRT displays cheaper than the 2500 USD 512 x 512 PLATO plasma displays. Nonetheless, the plasma displays' relatively large screen size and 1 inch thickness made them suitable for high-profile placement in lobbies and stock exchanges.

Burroughs Corporation, a maker of adding machines and computers, developed the Panaplex display in the early 1970s. The Panaplex display, generically referred to as a gas-discharge or gas-plasma display, uses the same technology as later plasma video displays, but began life as seven-segment display for use in adding machines. They became popular for their bright orange luminous look and found nearly ubiquitous use in cash registers, calculators, pinball machines, aircraft avionics such as radios, navigational instruments, and stormscopes; test equipment such as frequency counters and multimetres; and generally anything that previously used nixie tube or numitron displays with a high digit-count throughout the late 1970s and into the 1990s. These displays remained popular until LEDs gained popularity because of their low-current draw and module-flexibility, but are still found in some applications where their high-brightness is desired, such as pinball machines and avionics. Pinball displays started with six- and seven-digit seven-segment displays and later evolved into 16-segment alphanumeric displays, and later into 128x32 dot-matrix displays in 1990, which are still used today.

1983

In 1983, IBM introduced a 19-inch (48 cm) orange-on-black monochrome display (model 3290 'information panel') which was able to show up to four simultaneous IBM 3270 terminal sessions. Due to heavy competition from monochrome LCD's, in 1987 IBM planned to shut down its factory in upstate New York, the largest plasma plant in the world, in favor of manufacturing mainframe computers. Consequently, Larry Weber co-founded a startup company Plasmaco with Stephen Globus, as well as James Kehoe, who was the IBM plant manager, and bought the plant from IBM. Weber stayed in Urbana as CTO until 1990, then moved to upstate New York to work at Plasmaco.

1992

In 1992, Fujitsu introduced the world's first 21-inch (53 cm) full-colour display. It was a hybrid, the plasma display created at the University of Illinois at Urbana-Champaign and NHK Science & Technology Research Laboratories.

1994

In 1994, Weber demonstrated colour plasma technology at an industry convention in San Jose. Panasonic Corporation began a joint development project with Plasmaco, which led in 1996 to the purchase of Plasmaaco, its colour AC technology, and its American factory.

1997

In 1997, Fujitsu introduced the first 42-inch (107 cm) plasma display; it had 852x480 resolution and was progressively scanned. Also in 1997, Philips introduced a 42-inch (107 cm) display, with 852x480 resolution. It was the only plasma to be displayed to the retail public in 4 Sears locations in the US. The price was US$14,999 and included in-home installation. Later in 1997, Pioneer started selling their first plasma television to the public, and others followed.

2006–present

In late 2006, analysts noted that LCDs overtook plasmas, particularly in the 40-inch (1.0 m) and above segment where plasma had previously gained market share. Another industry trend is the consolidation of manufacturers of plasma displays, with around fifty brands available but only five manufacturers. In the first quarter of 2008 a comparison of worldwide TV sales breaks down to 22.1 million for direct-view CRT, 21.1 million for LCD, 2.8 million for Plasma, and 0.1 million for rear-projection.

Until the early 2000s, plasma displays were the most popular choice for HDTV flat panel display as they had many benefits over LCDs. Beyond plasma's deeper blacks, increased contrast, faster response time, greater colour spectrum, and wider viewing angle; they were also much bigger than LCDs, and it was believed that LCD technology was suited only to smaller sized televisions. However, improvements in VLSI fabrication technology have since narrowed the technological gap. The increased size, lower weight, falling prices, and often lower electrical power consumption of LCDs now make them competitive with plasma television sets.

Screen sizes have increased since the introduction of plasma displays. The largest plasma video display in the world at the 2008 Consumer Electronics Show in Las Vegas, Nevada, was a 150-inch (381 cm) unit manufactured by Matsushita Electric Industrial (Panasonic) standing 6 ft (180 cm) tall by 11 ft (330 cm) wide. At the 2010 Consumer Electronics Show in Las Vegas, Panasonic introduced their 152" 2160p 3D plasma. In 2010 Panasonic shipped 19.1 million plasma TV panels.

In 2010, the shipments of plasma TVs reached 18.2 million units globally.

Notable Plasma Display Manufacturers

- Panasonic Corporation (formerly Matsushita)
- Samsung Electronics
- LG Electronics
- Gradiente
- Lanix
- ProScan
- Sanyo
- Funai.

Notable Manufacturers that have Ceased Production of Plasma TV's

- Sony
- Hitachi Ltd.
- Philips
- Vizio
- Toshiba
- RCA
- NEC
- Pioneer Corporation (in March 2009 as well as television manufacturing altogether)
- Fujitsu.

Fluorescent Lamp

A fluorescent lamp or fluorescent tube is a gas-discharge lamp that uses electricity to excite mercury vapour. The excited mercury atoms produce short-wave ultraviolet light that then causes a phosphor to fluoresce, producing visible light. A fluorescent lamp converts electrical power into useful light more efficiently than an incandescent lamp. Lower energy cost typically offsets the higher initial cost of the lamp. The lamp fixture is more costly because it requires a ballast to regulate the current through the lamp.

While larger fluorescent lamps have been mostly used in commercial or institutional buildings, the compact fluorescent lamp is now available in the same popular sizes as incandescents and is used as an energy-saving alternative in homes.

The United States Environmental Protection Agency classifies fluorescent lamps as hazardous waste, and recommends that they be segregated from general waste for recycling or safe disposal.

Physical Discoveries

Fluorescence of certain rocks and other substances had been observed for hundreds of years before its nature was understood. By the middle of the 19th century, experimenters had observed a radiant glow emanating from partially evacuated glass vessels through which an electric current passed. One of the first to explain it was the Irish scientist Sir George Stokes from the University of Cambridge, who named the phenomenon "fluorescence" after fluorite, a mineral many of whose samples fluoresce strongly due to impurities. The explanation relied on the nature of electricity and light phenomena as developed by the British scientists Michael Faraday and James Clerk Maxwell in the 1840s.

Little more was done with this phenomenon until 1856 when a German glassblower named Heinrich Geissler created a mercury vacuum pump that evacuated a glass tube to an extent not previously possible. When an electrical current passed through a Geissler tube, a strong green glow on the walls of the tube at the cathode end could be observed. Because it produced some beautiful light effects, the Geissler tube was a popular source of amusement. More important, however, was its contribution to scientific research. One of the first scientists to experiment with a Geissler tube was Julius Plücker who systematically described in 1858 the luminescent effects that occurred in a Geissler tube. He also made the important observation that the glow in the tube shifted position when in proximity to an electromagnetic field. Alexandre Edmond Becquerel observed in 1859 that certain substances gave off light when they were placed in a Geissler tube. He went on to apply thin coatings of luminescent materials to the surfaces of these tubes. Fluorescence occurred, but the tubes were very inefficient and had a short operating life.

Inquiries that began with the Geissler tube continued as even better vacuums were produced. The most famous was the evacuated tube used for scientific research by William Crookes. That tube was evacuated by the highly effective mercury vacuum pump created by Hermann Sprengel. Research conducted by Crookes and others ultimately led to the discovery of the electron in 1897 by J. J. Thomson and X-rays in 1895 by Wilhelm Roentgen. But the Crookes tube, as it came to be known, produced little light because the vacuum in it was too good and thus lacked the trace amounts of gas that are needed for electrically stimulated luminescence.

Early Discharge Lamps

While Becquerel was primarily interested in conducting scientific research into fluorescence, Thomas Edison briefly pursued fluorescent lighting for its commercial potential. He invented a fluorescent lamp in 1896 that used a coating of calcium tungstate as the fluorescing substance, excited by X-rays, but although it received a patent in 1907, it was not put into production. As with a few other attempts to use Geissler tubes for illumination, it had a short operating life, and given the success of the incandescent light, Edison had little reason to pursue an alternative means of electrical illumination. Nikola Tesla made similar experiments in the 1890s, devising high frequency powered fluorescent bulbs that gave a bright greenish light, but as with Edison's devices, no commercial success was achieved.

Although Edison lost interest in fluorescent lighting, one of his former employees was able to create a gas-based lamp that achieved a measure of commercial success. In 1895 Daniel McFarlan Moore demonstrated lamps 2 to 3 meters (6.6 to 9.8 ft) in length that used carbon dioxide or nitrogen to emit white or pink light, respectively. As with future fluorescent lamps, they were considerably more complicated than an incandescent bulb.

After years of work, Moore was able to extend the operating life of the lamps by inventing an electromagnetically controlled valve that maintained a constant gas pressure within the tube. Although Moore's lamp was complicated, expensive to install, and required very high voltages, it was considerably more efficient than incandescent lamps, and it produced a more natural light than incandescent lamps. From 1904 onwards Moore's lighting system was installed in a number of stores and offices. Its success contributed to General Electric's motivation to improve the incandescent lamp, especially its filament. GE's efforts came to fruition with the invention of a tungsten-based filament. The extended lifespan of incandescent bulbs negated one of the key advantages of Moore's lamp, but GE purchased the relevant patents in 1912. These patents and the inventive efforts that supported them were to be of considerable value when the firm took up fluorescent lighting more than two decades later.

At about the same time that Moore was developing his lighting system, another American was creating a means of illumination that also can be seen as a precursor to the modern fluorescent lamp. This was the mercury-vapour lamp, invented by Peter Cooper Hewitt and patented in 1901 (US 682692; this patent number is universally misquoted as US889,692). Hewitt's lamp luminesced when an electric current was passed through mercury vapour at a low pressure. Unlike Moore's lamps, Hewitt's were manufactured in standardised sizes and operated at low voltages. The mercury-vapour lamp was superior to the incandescent lamps of the time in terms of energy efficiency, but the blue-green light it produced limited its applications. It was, however, used for photography and some industrial processes. Mercury vapour lamps continued to be developed at a slow pace, especially in Europe, and by the early 1930s they received limited use for large-scale illumination. Some of them employed fluorescent coatings, but these were primarily used for colour correction and not for enhanced light output. Mercury vapour lamps also anticipated the fluorescent lamp in their incorporation of a ballast to maintain a constant current.

Cooper-Hewitt had not been the first to use mercury vapour for illumination, as earlier efforts had been mounted by Way, Rapieff, Arons, and Bastian and Salisbury. Of particular importance was the mercury vapour lamp invented by Küch in Germany. This lamp used quartz in place of glass to allow higher operating temperatures, and hence greater efficiency. Although its light output relative to electrical consumption was better than other sources of light, the light it produced was similar to that of the Cooper-Hewitt lamp in that it lacked the red portion of the spectrum, making it unsuitable for ordinary lighting.

Neon Lamps

The next step in gas-based lighting took advantage of the luminescent qualities of neon, an inert gas that had been discovered in 1898 by isolation from the atmosphere. Neon glowed a brilliant red when used in Geissler tubes. By 1910, Georges Claude, a Frenchman who had developed a technology and a successful business for air liquefaction, was obtaining enough neon as a byproduct to support a neon lighting industry. While neon lighting was used around 1930 in France for general illumination, it was no more energy-efficient than conventional incandescent lighting.

Neon tube lighting, which also includes the use of argon and mercury vapour as alternate gases, came to be used primarily for eye-catching signs and advertisements. Neon lighting was relevant to the development of fluorescent lighting, however, as Claude's improved electrode (patented in 1915) overcame "sputtering", a major source of electrode degradation.

Sputtering occurred when ionised particles struck an electrode and tore off bits of metal. Although Claude's invention required electrodes with a lot of surface area, it showed that a major impediment to gas-based lighting could be overcome.

The development of the neon light also was significant for the last key element of the fluorescent lamp, its fluorescent coating. In 1926 Jacques Risler received a French patent for the application of fluorescent coatings to neon light tubes. The main use of these lamps, which can be considered the first commercially successful fluorescents, was for advertising, not general illumination. This, however, was not the first use of fluorescent coatings. As has been noted above, Edison used calcium tungstate for his unsuccessful lamp. Other efforts had been mounted, but all were plagued by low efficiency and various technical problems. Of particular importance was the invention in 1927 of a low-voltage "metal vapour lamp" by Friedrich Meyer, Hans-Joachim Spanner, and Edmund Germer, who were employees of a German firm in Berlin. A German patent was granted but the lamp never went into commercial production.

Commercialisation of Fluorescent Lamps

All the major features of fluorescent lighting were in place at the end of the 1920s. Decades of invention and development had provided the key components of fluorescent lamps: economically manufactured glass tubing, inert gases for filling the tubes, electrical ballasts, long-lasting electrodes, mercury vapour as a source of luminescence, effective means of producing a reliable electrical discharge, and fluorescent coatings that could be energised by ultraviolet light. At this point, intensive development was more important than basic research.

In 1934, Arthur Compton, a renowned physicist and GE consultant, reported to the GE lamp department on successful experiments with fluorescent lighting at General Electric Co., Ltd. in Great Britain (unrelated to General Electric in the United States). Stimulated by this report, and with all of the key elements available, a team led by George E. Inman built a prototype fluorescent lamp in 1934 at General Electric's Nela Park (Ohio) engineering laboratory. This was not a trivial exercise; as noted by Arthur A. Bright, "A great deal of experimentation had to be done on lamp sizes and shapes, cathode construction, gas pressures of both argon and mercury vapour, colours of fluorescent powders, methods of attaching them to the inside of the tube, and other details of the lamp and its auxiliaries before the new device was ready for the public."

In addition to having engineers and technicians along with facilities for R&D work on fluorescent lamps, General Electric controlled what it regarded as the key patents covering fluorescent lighting, including the patents originally issued to Hewitt, Moore, and Küch. More important than these was a patent covering an electrode that did not disintegrate at the gas pressures that ultimately were employed in fluorescent lamps. Albert W. Hull of GE's Schenectady Research Laboratory filed for a patent on this invention in 1927, which was issued in 1931.

While the Hull patent gave GE a basis for claiming legal rights over the fluorescent lamp, a few months after the lamp went into production the firm learned of a U.S. patent application that had been filed in 1927 for the aforementioned "metal vapour lamp" invented in Germany by Meyer, Spanner, and Germer. The patent application indicated that the lamp had been created as a superior means of producing ultraviolet light, but the application also contained a few statements referring to fluorescent illumination. Efforts to obtain a U.S. patent had met with numerous delays, but were it to be granted, the patent might have caused serious difficulties for GE. At first, GE sought to block the issuance of a patent by claiming that priority should go to one of their employees, Leroy J. Buttolph, who according to their claim had invented a fluorescent lamp in 1919 and whose patent application was still pending. GE also had filed a patent application in 1936 in Inman's name to cover the "improvements" wrought by his group. In 1939 GE decided that the claim of Meyer, Spanner, and Germer had some merit, and that in any event a long interference procedure was not in their best interest. They therefore dropped the Buttolph claim and paid $180,000 to acquire the Meyer, et al. application, which at that point was owned by a firm known as Electrons, Inc. The patent was duly awarded in December 1939. This patent, along with the Hull patent, put GE on what seemed to be firm legal ground, although it faced years of legal challenges from Sylvania Electric Products, Inc., which claimed infringement on patents that it held.

Even though the patent issue would not be completely resolved for many years, General Electric's strength in manufacturing and marketing the bulb gave it a pre-eminent position in the emerging fluorescent light market. Sales of "fluorescent lumiline lamps" commenced in 1938 when four different sizes of tubes were put on the market used in fixtures manufactured by three leading corporations, Lightolier, Artcraft Fluorescent Lighting Corporation, and Globe Lighting, two based in New York City. During the following year GE and Westing house publicised the new lights through exhibitions at the New York World's Fair and the Golden Gate International Exposition in San Francisco. Fluorescent lighting systems spread rapidly during World War II as wartime manufacturing intensified lighting demand. By 1951 more light was produced in the United States by fluorescent lamps than by incandescent lamps.

In the first years zinc orthosilicate with varying content of beryllium was used as greenish phosphor. Small additions of magnesium tungstate improved the blue part of the spectrum yielding acceptable white. After it was discovered that beryllium was toxic, halophosphate based phosphors took over.

Principles of Operation

The fundamental means for conversion of electrical energy into radiant energy in a fluorescent lamp relies on inelastic scattering of electrons. An incident electron collides with an atom in the gas. If the free electron has enough kinetic energy, it transfers energy to the atom's outer electron, causing that electron to temporarily jump up to a higher energy level. The collision is 'inelastic' because a loss of kinetic energy occurs.

This higher energy state is unstable, and the atom will emit an ultraviolet photon as the atom's electron reverts to a lower, more stable, energy level. Most of the photons that are released from the mercury atoms have wavelengths in the ultraviolet (UV) region of the spectrum, predominantly at wavelengths of 253.7 and 185 nanometres (nm). These are not visible to the human eye, so they must be converted into visible light. This is done by making use of fluorescence. Ultraviolet photons are absorbed by electrons in the atoms of the lamp's interior fluorescent coating, causing a similar energy jump, then drop, with emission of a further photon. The photon that is emitted from this second interaction has a lower energy than the one that caused it. The chemicals that make up the phosphor are chosen so that these emitted photons are at wavelengths visible to the human eye. The difference in energy between the absorbed ultra-violet photon and the emitted visible light photon goes toward heating up the phosphor coating.

When the light is turned on, the electric power heats up the cathode enough for it to emit electrons (thermionic emission). These electrons collide with and ionise noble gas atoms inside the bulb surrounding the filament to form a plasma by the process of impact ionisation. As a result of avalanche ionisation, the conductivity of the ionised gas rapidly rises, allowing higher currents to flow through the lamp.

The fill gas helps determine the operating electrical characteristics of the lamp, but does not give off light itself. The fill gas effectively increases the distance that electrons travel through the tube, which allows an electron a greater chance of interacting with a mercury atom. Argon atoms, excited to a metastable state by impact of an electron, can impart this energy to a neutral mercury atom and ionise it, described as the Penning effect. This has the benefit of lowering the breakdown and operating voltage of the lamp, compared to other possible fill gases such as krypton.

Construction

A fluorescent lamp tube is filled with a gas containing low pressure mercury vapour and argon, xenon, neon, or krypton. The pressure inside the lamp is around 0.3% of atmospheric pressure. The inner surface of the bulb is coated with a fluorescent (and often slightly phosphorescent) coating made of varying blends of metallic and rare-earth phosphor salts. The bulb's electrodes are typically made of coiled tungsten and usually referred to as cathodes because of their prime function of emitting electrons. For this, they are coated with a mixture of barium, strontium and calcium oxides chosen to have a low thermionic emission temperature.

Fluorescent lamp tubes are typically straight and range in length from about 100 millimetres (3.9 in) for miniature lamps, to 2.43 meters (8.0 ft) for high-output lamps. Some lamps have the tube bent into a circle, used for table lamps or other places where a more compact light source is desired. Larger U-shaped lamps are used to provide the same amount of light in a more compact area, and are used for special architectural purposes. Compact fluorescent lamps have several small-diameter tubes joined in a bundle of two, four, or six, or a small diameter tube coiled into a spiral, to provide a high amount of light output in little volume.

Light-emitting phosphors are applied as a paint-like coating to the inside of the tube. The organic solvents are allowed to evaporate, then the tube is heated to nearly the melting point of glass to drive off remaining organic compounds and fuse the coating to the lamp tube. Careful control of the grain size of the suspended phosphors is necessary; large grains, 35 micrometers or larger, lead to weak grainy coatings, whereas too many small particles 1 or 2 micrometers or smaller leads to poor light maintenance and efficiency. Most phosphors perform best with a particle size around 10 micrometers. The coating must be thick enough to capture all the ultraviolet light produced by the mercury arc, but not so thick that the phosphor coating absorbs too much visible light. The first phosphors were synthetic versions of naturally occurring fluorescent minerals, with small amounts of metals added as activators. Later other compounds were discovered, allowing differing colours of lamps to be made.

Electrical Aspects of Operation

Fluorescent lamps are negative differential resistance devices, so as more current flows through them, the electrical resistance of the fluorescent lamp drops, allowing even more current to flow. Connected directly to a constant-voltage power supply, a fluorescent lamp would rapidly self-destruct due to the uncontrolled current flow. To prevent this, fluorescent lamps must use an auxiliary device, a ballast, to regulate the current flow through the tube. The terminal voltage across an operating lamp varies depending on the arc current, tube diameter, temperature, and fill gas. A fixed part of the voltage drop is due to the electrodes. A general lighting service T12 1200-millimetre (48 in) lamp operates at 430 mA, with 100 volts drop. High output lamps operate at 800 mA, and some types operate up to 1500 mA. The power level varies from 33 to 82 watts per meter of tube length (10 to 25 W/ft) for T12 lamps.

The simplest ballast for alternating current use is an inductor placed in series, consisting of a winding on a laminated magnetic core. The inductance of this winding limits the flow of AC current. This type is still used, for example, in 120 volt operated desk lamps using relatively short lamps. Ballasts are rated for the size of lamp and power frequency. Where the mains voltage is insufficient to start long fluorescent lamps, the ballast is often a step-up auto transformer with substantial leakage inductance (so as to limit the current flow). Either form of inductive ballast may also include a capacitor for power factor correction.

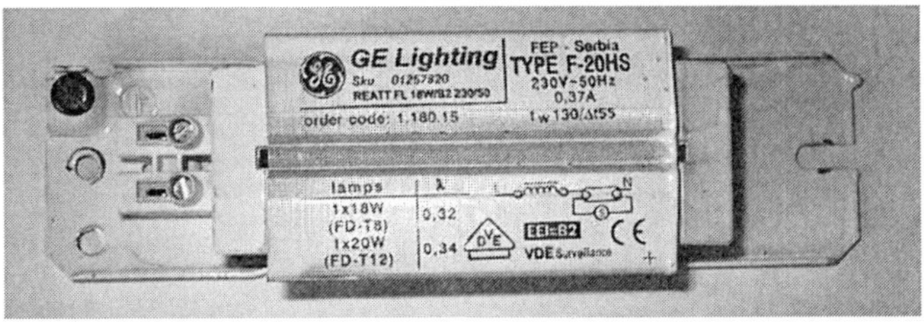

Figure 2: 230 V ballast for 18–20 W

Many different circuits have been used to operate fluorescent lamps. The choice of circuit is based on mains voltage, tube length, initial cost, long term cost, instant versus non-instant starting, temperature ranges and parts availability, etc. Fluorescent lamps can run directly from a DC supply of sufficient voltage to strike an arc. The ballast must be resistive, and would consume about as much power as the lamp. When operated from DC, the starting switch is often arranged to reverse the polarity of the supply to the lamp each time it is started; otherwise, the mercury accumulates at one end of the tube. Fluorescent lamps are (almost) never operated directly from DC for those reasons. Instead, an inverter converts the DC into AC and provides the current-limiting function as described below for electronic ballasts.

Effect of Temperature

The light output and performance of fluorescent lamps is critically affected by the temperature of the bulb wall and its effect on the partial pressure of mercury vapour within the lamp. Each lamp contains a small amount of mercury, which must vaporise to support the lamp current and generate light. At low temperatures the mercury is in the form of dispersed liquid droplets. As the lamp warms, more of the mercury is in vapour form. At higher temperatures, self-absorption in the vapour reduces the yield of UV and visible light. Since mercury condenses at the coolest spot in the lamp, careful design is required to maintain that spot at the optimum temperature, around 40 °C.

By using an amalgam with some other metal, the vapour pressure is reduced and the optimum temperature range extended upward;

however, the bulb wall "cold spot" temperature must still be controlled to prevent migration of the mercury out of the amalgam and condensing on the cold spot. Fluorescent lamps intended for higher output will have structural features such as a deformed tube or internal heat-sinks to control cold spot temperature and mercury distribution. Heavily loaded small lamps, such as compact fluorescent lamps, also include heat-sink areas in the tube to maintain mercury vapour pressure at the optimum value.

Losses

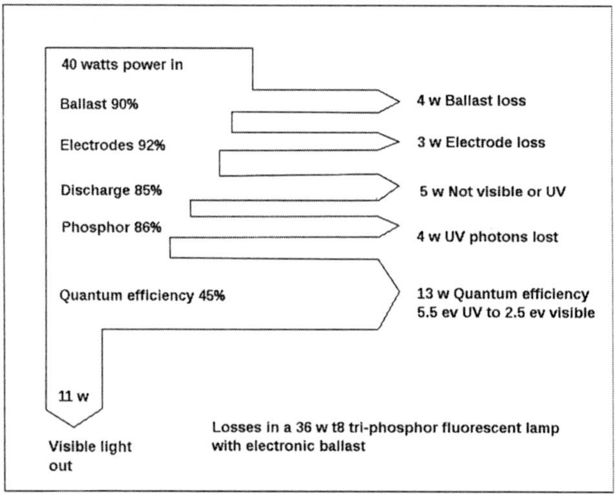

Figure 3: A Sankey diagram of energy losses in a fluorescent lamp. In modern designs, the biggest loss is the quantum efficiency of converting high-energy UV photons to lower-energy visible light photons.

The efficiency of fluorescent lighting owes much to the fact that low pressure mercury discharges emit about 65% of their total light in the 254 nm line (another 10–20% of the light is emitted in the 185 nm line). The UV light is absorbed by the bulb's fluorescent coating, which re-radiates the energy at longer wavelengths to emit visible light. The blend of phosphors controls the colour of the light, and along with the bulb's glass prevents the harmful UV light from escaping.

Only a fraction of the electrical energy input into a lamp is converted to useful light. The ballast dissipates some heat; electronic ballasts may be around 90% efficient. A fixed voltage drop occurs at the electrodes, which also produces heat. Some of the energy in the mercury vapour column is also dissipated, but about 85% is turned into visible and ultraviolet light.

Not all the UV energy striking the phosphor gets converted into visible light. In a modern lamp, for every 100 incident photons of UV impacting the phosphor, only 86 visible light photons are emitted (a quantum efficiency of 86%). The largest single loss in modern lamps is due to the lower energy of each photon of visible light, compared to the energy of the UV photons that generated them (a phenomenon called Stokes shift). Incident photons have an energy of 5.5 electron volts but produce visible light photons with energy around 2.5 electron volts, so only 45% of the UV energy is used; the rest is dissipated as heat. If a so-called "two-photon" phosphor could be developed, this would improve the efficiency but much research has not yet found such a system.

Cold Cathode Lamps

Most fluorescent lamps use electrodes that operate by thermionic emission, meaning they are operated at a high enough temperature for the electrode material (usually aided by a special coating) to emit electrons into the tube by heat.

However, there are also tubes that operate in cold cathode mode, whereby electrons are liberated into the tube only by the large potential difference (voltage) between the electrodes. This does not mean the electrodes are cold (indeed, they can be very hot), but it does mean they are operating below their thermionic emission temperature. Because cold cathode lamps have no thermionic emission coating to wear out they can have much longer lives than "hot cathode" thermionic emission tubes. This quality makes them desirable for maintenance-free long-life applications (such as LCD backlight displays). Sputtering of the electrode may still occur, but electrodes can be shaped (e.g. into an internal cylinder) to capture most of the sputtered material so it is not lost from the electrode.

Cold cathode lamps are generally less efficient than thermionic emission lamps because the cathode fall voltage is much higher. The increased fall voltage results in more power dissipation at tube ends, which does not contribute to light output. However, this is less significant with longer tubes. The increased power dissipation at tube ends also usually means cold cathode tubes have to be run at a lower loading than their thermionic emission equivalents. Given the higher tube voltage required anyway, these tubes can easily be made long, and even run as series strings. They are better suited for bending into special shapes for lettering and signage, and can also be instantly switched on or off.

Starting

The mercury atoms in the fluorescent tube must be ionised before the arc can "strike" within the tube. For small lamps, it does not take much voltage to strike the arc and starting the lamp presents no problem, but larger tubes require a substantial voltage (in the range of a thousand volts).

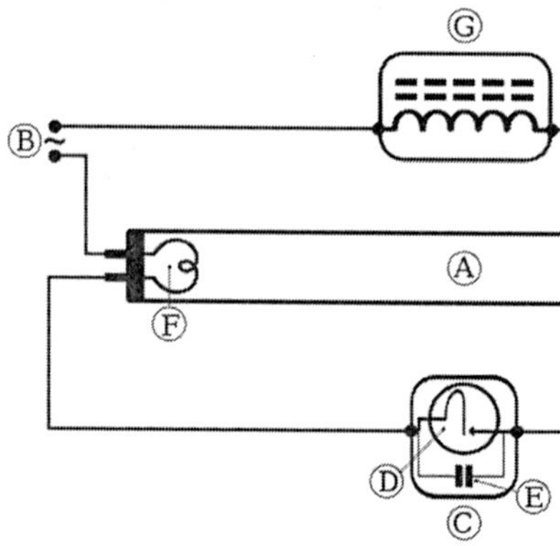

Figure 4: A *preheat* fluorescent lamp circuit using an automatic starting switch. A: Fluorescent tube, B: Power (+220 volts), C: Starter, D: Switch (bi-metallic thermostat), E: Capacitor, F: Filaments, G: Ballast

Switchstart or Preheat

This technique uses a combination filament–cathode at each end of the lamp in conjunction with a mechanical or automatic switch that initially connect the filaments in series with the ballast and thereby preheat the filaments prior to striking the arc. Note that in North America, this is referred to as *Preheat*. Elsewhere this is referred to as *Switchstart*.

These systems are standard equipment in 200–240 V countries (and for 100–120 V lamps up to about 30 watts), and generally use a glow starter. Before the 1960s, four-pin thermal starters and manual switches were also used. Electronic starters are also sometimes used with these electromagnetic ballast lamp fittings.

The automatic glow starter shown in the photograph to the left consists of a small gas-discharge lamp, containing neon and/or argon in parallel with a normally-open bi-metallic switch. It functions as a time-delay switch, turning the filaments of the fluorescent tube on for a few seconds to start it, then turning them off.

Figure 5: Electronic fluorescent lamp starters

When power is first applied to the circuit, aglow discharge will appear over the electrodes in the starter lamp. This glow discharge will heat the gas in the starter and cause the bi-metallic contact to bend towards the other contact. When the contacts touch, the two filaments of the fluorescent lamp and the ballast will effectively be switched in series to the supply voltage. The current through the filaments causes them to heat up and emit electrons into the tube gas by thermionic emission. In the starter, the touching contacts have extinguished the glow discharge, causing the gas to cool down again. The bi-metallic contact also cools down and starts to move back. Within a second or two the contacts separate, and the current through the filaments is interrupted, leaving the full line voltage applied between the filaments at the ends of the tube. The inductive kick from the interruption of current through the ballast provides the high voltage needed to start the lamp. The starter additionally has a capacitor wired in parallel to its gas-discharge tube, in order to prolong the contact life.

Once the tube is struck, the impinging main discharge then keeps the cathodes hot, permitting continued electron emission without the need for a separate current to heat the filaments. The starter does not close again because the voltage across the lit tube is insufficient to start a glow discharge in the starter. The starter is reliable because if the tube does not strike, the glow discharge recurs and the starter cycles again. Glow starters will often cycle a few times before the filaments get hot enough to start the tube, which causes an undesirable flashing during starting (The older thermal starters behaved better in this respect).

With automated starters such as glow starters, a failing tube will cycle endlessly, flashing as the lamp quickly goes out because the emission mix is insufficient to keep the lamp current high enough to keep the glow starter open. This causes flickering, and runs the ballast at above design temperature. Some more advanced starters time out in this situation, and do not attempt repeated starts until power is reset. Some older systems used a thermal over-current trip to detect repeated starting attempts. These require manual reset. The switch contacts in glow starters eventually wear out and fail, so the starter is manufactured as a separate unit that plugs into a socket in the lamp's housing, so it can be replaced.

Electronic starters use a more complex method to preheat the cathodes of a fluorescent lamp. Electronic starters are made in the same physical case as glow starters for direct replacement. They commonly use a specially designed semiconductor switch. They are programmed with a predefined preheat time to ensure that the cathodes are fully heated and reduce the amount of sputtered emission mix to prolong the life of the lamp; typically it is claimed that the life of a lamp frequently switched on, as in domestic use, is prolonged by a factor of 3 to 4 times. Start time is typically 1 to 4 seconds. Electronic starters contain a series of capacitors that are capable of producing a high voltage pulse of electricity across the lamp to ensure that it strikes correctly. Electronic starters only attempt to start a lamp for a short time when power is initially applied and will not repeatedly attempt to restrike a lamp that is dead and cannot sustain an arc; some will automatically shut down a failed lamp. This eliminates the re-striking of a lamp and the continuous flickering on and off of a failing lamp with a glow starter. Some fast-start electronic starters can strike the fluorescent tube within 0.3 seconds.

Instant Start

Another type of tube doesn't have filaments to start it at all. *Instant start* fluorescent tubes simply use a high enough voltage to break down the gas and mercury column and thereby start arc conduction. These tubes can be identified by a single pin at each end of the tube. The lamp holders have a "disconnect" socket at the low-voltage end which disconnects the ballast when the tube is removed, to prevent electric shock. Low-cost lighting fixtures with an integrated electronic ballast use instant start on preheat lamps, even if it reduces the lamp lifespan.

Rapid Start

Newer *rapid start* ballast designs provide filament power windings within the ballast; these rapidly and continuously warm the filaments/

cathodes using low-voltage AC. No inductive voltage spike is produced for starting, so the lamps must be mounted near a grounded (earthed) reflector to allow the glow discharge to propagate through the tube and initiate the arc discharge. In some lamps a grounded "starting aid" strip is attached to the outside of the lamp glass.

Quick-Start

Quick-start ballasts use a small auto-transformer to heat the filaments when power is first applied. When an arc strikes, the filament heating power is reduced and the tube will start within half a second. The auto-transformer is either combined with the ballast or may be a separate unit. Tubes need to be mounted near an earthed metal reflector in order for them to strike. Quick-start ballasts were more common in commercial installations because of lower maintenance as no starter switches need to be replaced. They are also used in domestic installations due to the virtually instant start. Quick-start ballasts are only used on 240 V circuits and are designed for use with the older, less-efficient T12 tubes, T8 retrofits will not start when used with quick-start ballasts.

Semi-resonant Start

Semi-resonant start was invented by Thorn Lighting for use with T12 fluorescent tubes. This method uses a double wound transformer and a capacitor. With no arc current, the transformer and capacitor resonate at mains frequency and generate about twice mains voltage across the tube, and a small electrode heating current. This tube voltage is too low to strike the arc with cold electrodes, but as the electrodes heat up to thermionic emission temperature, the tube striking voltage reduces below that of the ringing voltage, and the arc strikes. As the electrodes heat, the lamp slowly, over 3–5 seconds, reaches full brightness. As the arc current increases and tube voltage drops, the circuit provides current limiting.

Semi-resonant start was mainly used in commercial installations because of their higher initial cost. There are no starter switches to be replaced and cathode damage is reduced during starting. Due to the high open circuit tube voltage, this starting method was particularly good for starting tubes in cold locations. Additionally, the circuit power factor is almost 1.0, and no additional power factor correction is needed in the lighting installation. As the design requires that twice the mains voltage must be lower than the cold-cathode striking voltage (or the tubes would erroneously instant-start), this design can only be used with 5 ft and longer tubes on 240 V mains. Semi-resonant start fixtures are generally incompatible with energy saving T8 retrofit tubes, because such tubes have a higher starting voltage than T12 lamps and may not start reliably, especially in low temperatures. Recent proposals in some countries to phase out T12 tubes will reduce the application of this starting method.

Programmed Start

This is used with electronic ballasts shown below. A programmed-start ballast is a more advanced version of rapid start. This ballast applies power to the filaments first, then after a short delay to allow the cathodes to preheat, applies voltage to the lamps to strike an arc.

This ballast gives the best life and most starts from lamps, and so is preferred for applications with very frequent power cycling such as vision examination rooms and restrooms with a motion detector switch.

Electronic Ballasts

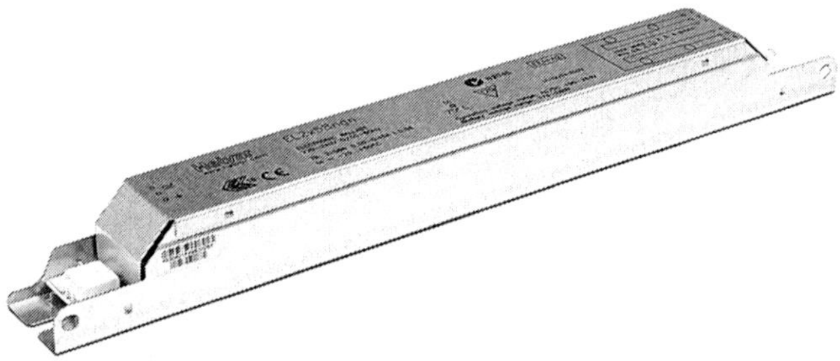

Figure 6: Electronic ballast for fluorescent lamp, 2x58W

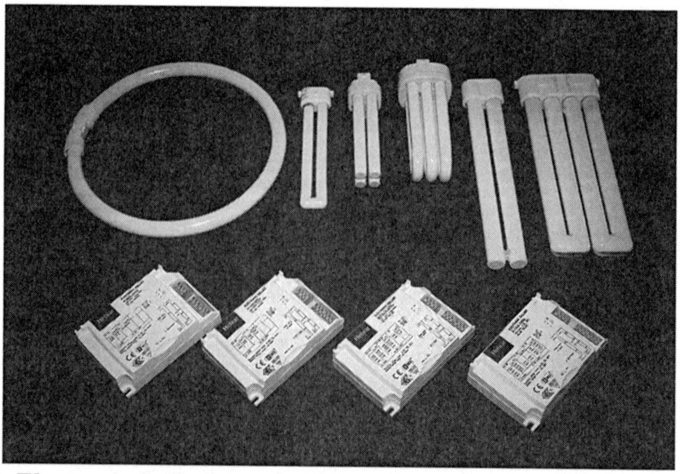

Figure 7: Electronic ballasts and different compact fluorescent lamps

Electronic ballasts employ transistors to alter mains voltage frequency into high-frequency AC while also regulating the current flow in the lamp. Some still use an inductance to limit the current, but the higher frequency allows a much smaller inductance to be used. Others use a capacitor-transistor combination to replace the inductor, since a transistor and capacitor working together can perfectly simulate the action of an inductor. These ballasts take advantage of the higher efficacy of lamps operated with higher-frequency current. Efficacy of a fluorescent lamp rises by almost 10% at a frequency of 10 kHz, compared to efficacy at normal power frequency. When the AC period is shorter than the relaxation time to de-ionise mercury atoms in the discharge column, the discharge stays closer to optimum operating condition. Electronic ballasts typically work in rapid start or instant start mode. Electronic ballasts are commonly supplied with AC power, which is internally converted to DC and then back to a variable frequency AC waveform. Depending upon the capacitance and the quality of constant-current pulse-width modulation, this can largely eliminate modulation at 100 or 120 Hz.

Low cost ballasts mostly contain only a simple oscillator and series resonant LC circuit. When turned on, the oscillator starts, and the LC circuit charges. After a short time the voltage across the lamp reaches about 1 kV and the lamp ignites. The process is too fast to preheat the cathodes, so the lamp instant-starts in cold cathode mode. The cathode filaments are still used for protection of the ballast from overheating if the lamp does not ignite. A few manufacturers use positive temperature coefficient (PTC) thermistors to disable instant starting and give some time to preheat the filaments.

More complex electronic ballasts use programmed start. The output AC frequency is started above the resonance frequency of the output circuit of the ballast; and after the filaments are heated, the frequency is rapidly decreased. If the frequency approaches the resonant frequency of the ballast, the output voltage will increase so much that the lamp will ignite. If the lamp does not ignite, an electronic circuit stops the operation of the ballast.

Many electronic ballasts are controlled by a microcontroller or similar, and these are sometimes called digital ballasts. Digital ballasts can apply quite complex logic to lamp starting and operation. This enables functions such as testing for broken electrodes and missing tubes before attempting to start, auto detect tube replacement, and auto detection of tube type, such that a single ballast can be used with several different tubes, even those that operate at different arc currents, etc. Once such fine grained control over the starting and arc current is achievable, features such as dimming, and having the ballast maintain a constant light level against changing sunlight contribution are all easily included in the embedded microcontroller software, and can be found in various manufacturers' products.

Since introduction in the 1990s, high frequency ballasts have been used in general lighting fixtures with either rapid start or pre-heat lamps. These ballasts convert the incoming power to an output frequency in excess of 20 kHz. This increases lamp efficiency. These are used in several applications, including new generation tanning lamp systems, whereby a 100 watt lamp (e.g., F71T12BP) can be lit using 65 to 70 watts of actual power while obtaining the same luminous flux (measured in lumens) as magnetic ballasts. These ballasts operate with voltages that can be almost 600 volts, requiring some consideration in housing design, and can cause a minor limitation in the length of the wire leads from the ballast to the lamp ends.

End of Life

The end of life failure mode for fluorescent lamps varies depending on how they are used and their control gear type. Often the light will turn pink with black burns on the ends of the bulb due to sputtering of emission mix. The lamp may also flicker at a noticeable rate. More information about other normal tube failure modes including the above are as follows:

Emission Mix

The "emission mix" on the tube filaments/cathodes is necessary to enable electrons to pass into the gas viathermionic emission at the tube operating voltages used. The mix is slowly sputtered off by bombardment with electrons and mercury ions during operation, but a larger amount is sputtered off each time the tube is started with cold cathodes. The method of starting the lamp has a significant impact on this. Lamps operated for typically less than 3 hours each switch-on will normally run out of the emission mix before other parts of the lamp fail. The sputtered emission mix forms the dark marks at the tube ends seen in old tubes. When all the emission mix is gone, the cathode cannot pass sufficient electrons into the gas fill to maintain the discharge at the designed tube operating voltage. Ideally, the control gear should shut down the tube when this happens. However, some control gear will provide sufficient increased voltage to continue operating the tube in cold cathode mode, which will cause overheating of the tube end and rapid disintegration of the electrodes (filament goes open-circuit) and filament support wires until they are completely gone or the glass cracks, wrecking the low pressure gas fill and stopping the gas discharge.

Ballast Electronics

This may occur in compact fluorescent lamps with integral electrical ballasts or in linear lamps. Ballast electronics failure is a somewhat random process that follows the standard failure profile for any electronic device. There is an initial small peak of early failures, followed by a drop and steady increase over lamp life. Life of electronics is heavily dependent on operating temperature—it typically halves for each 10 °C temperature rise. The quoted average life of a lamp is usually at 25 °C ambient (this may vary by country). The average life of the electronics at this temperature is normally greater than this, so at this temperature, not many lamps will fail due to failure of the electronics. In some fittings, the ambient temperature could be well above this, in which case failure of the electronics may become the predominant failure mechanism. Similarly, running a compact fluorescent lamp base-up will result in hotter electronics, which can cause shorter average life (particularly with higher power rated ones). Electronic ballasts should be designed to shut down the tube when the emission mix runs out as described above. In the case of integral electronic ballasts, since they never have to work again, this is sometimes done by having them deliberately burn out some component to permanently cease operation.

In most CFLs the filaments are connected in series, with a small capacitor between them. The discharge, once lit, is in parallel to the capacitor and presents a lower-resistance path, effectively shorting the capacitor out. One of the most common failure modes of cheap lamps is caused by underrating this capacitor (using lower-voltage, lower-cost part), which is very stressed during operation, leading to its premature failure.

Phosphor

The phosphor drops off in efficiency during use. By around 25,000 operating hours, it will typically be half the brightness of a new lamp (although some manufacturers claim much longer half-lives for their lamps). Lamps that do not suffer failures of the emission mix or integral ballast electronics will eventually develop this failure mode. They still work, but have become dim and inefficient. The process is slow, and often only becomes obvious when a new lamp is operating next to an old one.

Loss of Mercury

Like in all mercury-based gas-filled tubes, mercury is slowly absorbed into glass, phosphor, and tube electrodes throughout the lamp life, where it can no longer function. Newer lamps now have just enough mercury to last the expected life of the lamp. Loss of mercury will take over from failure of the phosphor in some lamps. The failure symptoms are similar, except loss of mercury initially causes an extended run-up time to full light output, and finally causes the lamp to glow a dim pink when the mercury runs out and the argon base gas takes over as the primary discharge.

Subjecting the tube to asymmetric waveforms, where the total current flow through the tube does not cancel out and the tube effectively operates under a DC bias, causes asymmetric distribution of mercury ions along the tube due to cataphoresis. The localised depletion of mercury vapour pressure manifests as pink luminescence of the base gas in the vicinity of one of the electrodes, and the operating lifetime of the lamp may be dramatically shortened. This can be an issue with some poorly designed inverters.

The same effect can be observed with new tubes. Mercury is present in the form of an amalgam and takes some time to be liberated in sufficient amount. New lamps may initially glow pink for several seconds after startup. This period is minimised after about 100 hours of operation.

Burned Filaments

The filaments can burn at the end of the lamp's lifetime, opening the circuit and losing the capability to heat up. Both filaments lose function as they are connected in series, with just a simple switch start circuit a broken filament will render the lamp completely useless. Filaments rarely burn or fail open circuit unless the filament becomes depleted of emitter and the control gear is able to supply a high enough voltage across the tube to operate it in cold cathode mode. Some digital electronic ballasts are capable of detecting broken filaments and can still strike an arc with one or both filaments broken providing there is still sufficient emitter. A broken filament in a bulb attached to a magnetic ballast often causes both bulbs to burn out or flicker.

Phosphors and the Spectrum of Emitted Light

The spectrum of light emitted from a fluorescent lamp is the combination of light directly emitted by the mercury vapour, and light emitted by the phosphorescent coating. The spectral lines from the mercury emission and the phosphorescence effect give a combined spectral distribution of light that is different from those produced by incandescent sources. The relative intensity of light emitted in each narrow band of wavelengths over the visible spectrum is in different proportions compared to that of an incandescent source. Coloured objects are perceived differently under light sources with differing spectral distributions. For example, some people find the colour rendition produced by some fluorescent lamps to be harsh and displeasing. A healthy person can sometimes appear to have an unhealthy skin tone under fluorescent lighting. The extent to which this phenomenon occurs is related to the light's spectral composition, and may be gauged by its colour rendering index (CRI).

Colour Temperature

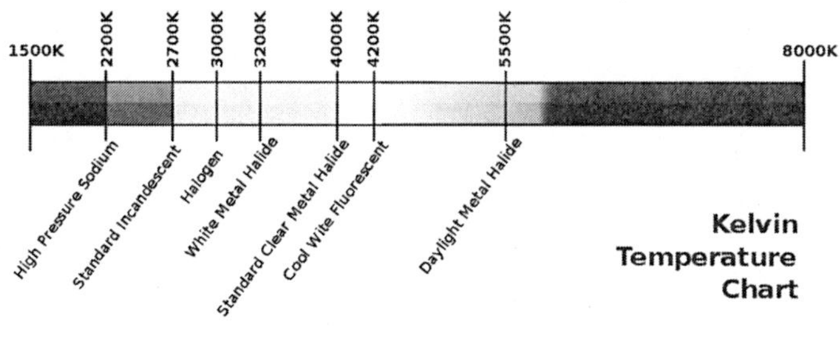

Figure 8: The colour temperature of different electric lamps

Correlated colour temperature (CCT) is a measure of the "shade" of whiteness of a light source, again by comparison with a blackbody. Typical incandescent lighting is 2700 K, which is yellowish-white. Halogen lighting is 3000 K. Fluorescent lamps are manufactured to a chosen CCT by altering the mixture of phosphors inside the tube. Warm-white fluorescents have CCT of 2700 K and are popular for residential lighting. Neutral-white fluorescents have a CCT of 3000 K or 3500 K. Cool-white fluorescents have a CCT of 4100 K and are popular for office lighting. Daylight fluorescents have a CCT of 5000 K to 6500 K, which is bluish-white.

High CCT lighting generally requires higher light levels. At dimmer illumination levels, the human eye perceives lower colour temperatures as more pleasant, as related through the Kruith of curve. So, a dim

2700 K incandescent lamp appears comfortable and a bright 5000 K lamp also appears natural, but a dim 5000 K fluorescent lamp appears too pale. Daylight-type fluorescents look natural only if they are very bright.

Colour Rendering Index

Colour rendering index (CRI) is a measure of how well colours can be perceived using light from a source, relative to light from a reference source such as daylight or a blackbody of the same colour temperature. By definition, an incandescent lamp has a CRI of 100. Real-life fluorescent tubes achieve CRIs of anywhere from 50 to 99. Fluorescent lamps with low CRI have phosphors that emit too little red light. Skin appears less pink, and hence "unhealthy" compared with incandescent lighting. Coloured objects appear muted. For example, a low CRI 6800 K halophosphate tube (an extreme example) will make reds appear dull red or even brown. Since the eye is relatively less efficient at detecting red light, an improvement in colour rendering index, with increased energy in the red part of the spectrum, may reduce the overall luminous efficacy.

Lighting arrangements use fluorescent tubes in an assortment of tints of white. Sometimes this is because of the lack of appreciation for the difference or importance of differing tube types. Mixing tube types within fittings can improve the colour reproduction of lower quality tubes.

Phosphor Composition

Some of the least pleasant light comes from tubes containing the older, halophosphate-type phosphors (chemical formula $Ca_5(PO_4)_3(F, Cl):Sb^{3+}, Mn^{2+}$). This phosphor mainly emits yellow and blue light, and relatively little green and red. In the absence of a reference, this mixture appears white to the eye, but the light has an incomplete spectrum. The CRI of such lamps is around 60.

Since the 1990s, higher quality fluorescent lamps use either a higher CRI halophosphate coating, or a *triphosphor* mixture, based on europium and terbium ions, that have emission bands more evenly distributed over the spectrum of visible light. High CRI halophosphate and triphosphor tubes give a more natural colour reproduction to the human eye. The CRI of such lamps is typically 82–100.

Applications

Fluorescent light bulbs come in many shapes and sizes. The compact fluorescent light bulb (CFL) is becoming more popular.

Many compact fluorescent lamps integrate the auxiliary electronics into the base of the lamp, allowing them to fit into a regular light bulb socket.

In US residences, fluorescent lamps are mostly found in kitchens, basements, or garages, but schools and businesses find the cost savings of fluorescent lamps to be significant and rarely use incandescent lights. Tax incentives and building code result in higher use in places such as California.

In other countries, residential use of fluorescent lighting varies depending on the price of energy, financial and environmental concerns of the local population, and acceptability of the light output. In East and Southeast Asia it is very rare to see incandescent bulbs in buildings anywhere. Some countries are encouraging the phase-out of incandescent light bulbs and substitution of incandescent lamps with fluorescent lamps or other types of energy-efficient lamps.

In addition to general lighting, special fluorescent lights are often used in stage lighting for film and video production. They are cooler than traditional halogen light sources, and use high-frequency ballasts to prevent video flickering and high colour-rendition index bulbs to approximate daylight colour temperatures.

Advantages

Luminous Efficacy: Fluorescent lamps convert more of the input power to visible light than incandescent lamps. A typical 100 watt tungsten filament incandescent lamp may convert only 2% of its power input to visible white light, whereas typical fluorescent lamps convert about 22% of the power input to visible white light.

The efficacy of fluorescent tubes ranges from about 16 lumens per watt for a 4 watt tube with an ordinary ballast to over 100 lumens per watt with a modern electronic ballast, commonly averaging 50 to 67 lm/W overall. Most compact fluorescents above 13 watts with integral electronic ballasts achieve about 60 lm/W. Lamps are rated by lumens after 100 hours of operation. For a given fluorescent tube, a high-frequency electronic ballast gives about a 10% efficacy improvement over an inductive ballast. It is necessary to include the ballast loss when evaluating the efficacy of a fluorescent lamp system; this can be about 25% of the lamp power with magnetic ballasts, and around 10% with electronic ballasts.

Fluorescent lamp efficacy is dependent on lamp temperature at the coldest part of the lamp. In T8 lamps this is in the centre of the tube. In T5 lamps this is at the end of the tube with the text stamped on it. The ideal temperature for a T8 lamp is 25 °C (77 °F) while the T5 lamp is ideally at 35 °C (95 °F).

Life

Typically a fluorescent lamp will last between 10 to 20 times as long as an equivalent incandescent lamp when operated several hours at a time. The higher initial cost of a fluorescent lamp is usually more than compensated for by lower energy consumption over its life. The longer life may also reduce lamp replacement costs, providing additional saving especially where labour is costly. Therefore they are widely used by businesses and institutions, but not as much by households.

Lower Luminosity

Compared with an incandescent lamp, a fluorescent tube is a more diffuse and physically larger light source. In suitably designed lamps, light can be more evenly distributed without point source of glare such as seen from an undiffused incandescent filament; the lamp is large compared to the typical distance between lamp and illuminated surfaces.

Lower Heat

About two-thirds to three-quarters less heat is given off by fluorescent lamps compared to an equivalent installation of incandescent lamps. This greatly reduces the size, cost, and energy consumption.

Disadvantages

Frequent Switching: If the lamp is installed where it is frequently switched on and off, it will age rapidly. Under extreme conditions, its lifespan may be much shorter than a cheap incandescent lamp. Each start cycle slightly erodes the electron-emitting surface of the cathodes; when all the emission material is gone, the lamp cannot start with the available ballast voltage. Fixtures intended for flashing of lights (such as for advertising) will use a ballast that maintains cathode temperature when the arc is off, preserving the life of the lamp.

The extra energy used to start a fluorescent lamp is equivalent to a few seconds of normal operation; it is more energy-efficient to switch off lamps when not required for several minutes.

Health and Safety Issues

If a fluorescent lamp is broken, a very small amount of mercury can contaminate the surrounding environment. About 99% of the mercury is typically contained in the phosphor, especially on lamps that are near the end of their life. The broken glass is usually considered a greater hazard than the small amount of spilled mercury. The EPA recommends airing out the location of a fluorescent tube break and using wet paper towels to help pick up the broken glass and fine particles. Any glass and used towels should be disposed of in a sealed plastic bag. Vacuum cleaners can cause the particles to become airborne, and should not be used.

Ultraviolet Emission

Fluorescent lamps emit a small amount of ultraviolet (UV) light. A 1993 study in the US found that UV exposure from sitting under fluorescent lights for eight hours is equivalent to only one minute of sun exposure. Very sensitive individuals may experience a variety of health problems relating to light sensitivity that is aggravated by artificial lighting.

UV light can affect sensitive paintings, especially watercolours and many textiles. Valuable art work must be protected from light by additional glass or transparent acrylic sheets put between the lamp(s) and the painting.

Ballast

Fluorescent lamps require a ballast to stabilise the current through the lamp, and to provide the initial striking voltage required to start the arc discharge. This increases the cost of fluorescent light fixtures, though often one ballast is shared between two or more lamps. Electromagnetic ballasts with a minor fault can produce an audible humming or buzzing noise. Magnetic ballasts are usually filled with atar-like potting compound to reduce emitted noise. Hum is eliminated in lamps with a high-frequency electronic ballast. Energy lost in magnetic ballasts can be significant, on the order of 10% of lamp input power. Electronic ballasts reduce this loss. Small lamps may use an incandescent lamp as a ballast if the supply voltage is high enough to allow the lamp to start.

Power Quality and Radio Interference

Simple inductive fluorescent lamp ballasts have a power factor of less than unity. Inductive ballasts include power factor correction capacitors. Simple electronic ballasts may also have low power factor due to their rectifier input stage.

Fluorescent lamps are a non-linear load and generate harmonic currents in the electrical power supply. The arc within the lamp may generate radio frequency noise, which can be conducted through power wiring. Suppression of radio interference is possible. Very good suppression is possible, but adds to the cost of the fluorescent fixtures.

Operating Temperature

Fluorescent lamps operate best around room temperature. At much lower or higher temperatures, efficiency decreases. At below-freezing temperatures standard lamps may not start. Special lamps may be needed for reliable service outdoors in cold weather. In applications such as road and railway signalling, fluorescent lamps which do not generate as much heat as incandescent lamps may not melt snow and ice build up around the lamp, leading to reduced visibility.

Lamp Shape

Fluorescent tubes are long, low-luminance sources compared with high pressure arc lamps and incandescent lamps. However, low luminous intensity of the emitting surface is useful because it reduces glare. Lamp fixture design must control light from a long tube instead of a compact globe.

The compact fluorescent lamp (CFL) replaces regular incandescent bulbs. However, some CFLs will not fit some lamps, because the harp (heavy wire shade support bracket) is shaped for the narrow neck of an incandescent lamp, while CFLs tend to have a wide housing for their electronic ballast close to the bulb's base.

Flicker Problems

Fluorescent lamps using a magnetic mains frequency ballast do not give out a steady light; instead, they flicker at twice the supply frequency. This results in fluctuations not only with light output but colour temperature as well, which may pose problems for photography and people who are sensitive to the flicker. Even among persons not sensitive to light flicker, a stroboscopic effect can be noticed, where something spinning at just the right speed may appear stationary if illuminated solely by a single fluorescent lamp. This effect is eliminated by paired lamps operating on a lead-lag ballast. Unlike a true strobe lamp, the light level drops in appreciable time and so substantial "blurring" of the moving part would be evident.

In some circumstances, fluorescent lamps operated at mains frequency can also produce flicker at the mains frequency (50 or 60 Hz) itself, which is noticeable by more people. This can happen in the last few hours of tube life when the cathode emission coating at one end has almost run out, and that cathode starts having difficulty emitting enough electrons into the gas fill, resulting in slight rectification and hence uneven light output in positive and negative going mains cycles. Mains frequency flicker can also sometimes be emitted from the very ends of the tubes, if each tube electrode produces a slightly different light output pattern on each half-cycle. Flicker at mains frequency is more noticeable in the peripheral vision than it is when viewed directly, as is all flicker (since the peripheral vision is faster—has a higher critical frequency—than the central vision).

New fluorescent lamps may show a twisting spiral pattern of light in a part of the lamp. This effect is due to loose cathode material and usually disappears after a few hours of operation.

Electromagnetic ballasts may also cause problems for video recording as there can be a "beat effect" between the periodic reading of a camera's sensor and the fluctuations in intensity of the fluorescent lamp. Fluorescent lamps using high-frequency electronic ballasts do not produce visible light flicker, since above about 5 kHz, the excited electron state half-life is longer than a half cycle, and light production becomes continuous. Operating frequencies of electronic ballasts are selected to avoid interference with infrared remote controls. Poor quality (or failing) electronic ballasts may have insufficient reservoir capacitance or have poor regulation, thereby producing considerable 100/120 Hz modulation of the light.

Dimming

Fluorescent light fixtures cannot be connected to dimmer switches intended for incandescent lamps. Two effects are responsible for this: the waveform of the voltage emitted by a standard phase-control dimmer interacts badly with many ballasts, and it becomes difficult to sustain an arc in the fluorescent tube at low power levels. Dimming installations require a compatible dimming ballast. These systems keep the cathodes of the fluorescent tube fully heated even as the arc current is reduced, promoting easy thermionic emission of electrons into the arc stream. CFLs are available that work in conjunction with a suitable dimmer.

Disposal and Recycling

The disposal of phosphor and particularly the toxic mercury in the tubes is an environmental issue. Governmental regulations in many areas require special disposal of fluorescent lamps separate from general and household wastes. For large commercial or industrial users of fluorescent lights, recycling services are available in many nations, and may be required by regulation. In some areas, recycling is also available to consumers.

Lamp Sizes and Designations

Systematic nomenclature identifies mass-market lamps as to general shape, power rating, length, colour, and other electrical and illuminating characteristics.

Other Fluorescent Lamps

Black Lights

Blacklights are a subset of fluorescent lamps that are used to provide near ultraviolet light (at about 360 nm wavelength). They are built in the same fashion as conventional fluorescent lamps but the glass tube is coated with a phosphor that converts the short-wave UV within the tube to long-wave UV rather than to visible light. They are used to provoke fluorescence (to provide dramatic effects using blacklight paint and to detect materials such as urine and certain dyes that would be invisible in visible light) as well as to attract insects to bug zappers.

So-called *blacklite blue* lamps are also made from more expensive deep purple glass known as Wood's glass rather than clear glass. The deep purple glass filters out most of the visible colours of light directly emitted by the mercury-vapour discharge, producing proportionally less visible light compared with UV light. This allows UV-induced fluorescence to be seen more easily (thereby allowing blacklight posters to seem much more dramatic). The blacklight lamps used in bug zappers do not require this refinement so it is usually omitted in the interest of cost; they are called simply *blacklite* (and not blacklite blue).

Tanning Lamps

The lamps used in tanning beds contain a different phosphor blend (typically 3 to 5 or more phosphors) that emits both UVA and UVB, provoking a tanning response in most human skin. Typically, the output is rated as 3% to 10% UVB (5% most typical) with the remaining UV as UVA. These are mainly F71, F72 or F73 HO (100 W) lamps, although 160 W VHO are somewhat common. One common phosphor used in these lamps is lead-activated barium disilicate, but a europium-activated strontium fluoroborate is also used. Early lamps used thallium as an activator, but emissions of thallium during manufacture were toxic.

Grow Lamps

Grow lamps contain phosphor blends that encourage photosynthesis, growth, and/or flowering in plants, algae, photosynthetic bacteria, and other light-dependent organisms. These often emit light in the red and blue colour range, which is absorbed by chlorophyll and used for photosynthesis in plants.

Infrared Lamps

Lamps can be made with a lithium metaluminate phosphor activated with iron. This phosphor has peak emissions between 675 and 875 nanometres, with lesser emissions in the deep red part of the visible spectrum.

Bilirubin Lamps

Deep blue light generated from a europium-activated phosphor is used in the light therapy treatment of jaundice; light of this colour penetrates skin and helps in the breakup of excess bilirubin.

Germicidal Lamps

Germicidal lamps depend on the property that UV light kills most germs. Germicidal lamps contain no phosphor at all (making them gas discharge lamps rather than fluorescent) and their tubes are made of fused quartz that is transparent to the UV light emitted by the mercury discharge. The UV emitted by these tubes will kill germs and ionise oxygen to ozone. In addition it can cause eye and skin damage and should not be used or observed without eye and skin protection. Besides their uses to kill germs and create ozone, they are sometimes used by geologists to identify certain species of minerals by the colour of their fluorescence. When used in this fashion, they are fitted with filters in the same way as blacklight-blue lamps are; the filter passes the short-wave UV and blocks the visible light produced by the mercury discharge. They are also used in some EPROM erasers.

Germicidal lamps have designations beginning with G (meaning 'Germicidal'), rather than F, for example G30T8 for a 30-watt, 1-inch (2.5 cm) diameter, 36-inch (91 cm) long germicidal lamp (as opposed to an F30T8, which would be the fluorescent lamp of the same size and rating).

Electrodeless Lamps

Electrodeless induction lamps are fluorescent lamps without internal electrodes. They have been commercially available since 1990. A current is induced into the gas column using electromagnetic induction. Because the electrodes are usually the life-limiting element of fluorescent lamps, such electrodeless lamps can have a very long service life, although they also have a higher purchase price.

Cold-cathode Fluorescent Lamps (CCFL)

Cold-cathode fluorescent lamps are used as backlighting for LCD displays in personal computer and TV monitors. They are also popular with computer case modders in recent years.

Science Demonstrations

Fluorescent lamps can be illuminated by means other than a proper electrical connection. These other methods, however, result in very dim or very short-lived illumination, and so are seen mostly in science demonstrations. Static electricity or a Van de Graaff generator will cause a lamp to flash momentarily as it discharges a high voltage capacitance. A Tesla coil will pass high frequency current through the tube, and since it has a high voltage as well, the gases within the tube will ionise and emit light. Capacitive coupling with high-voltage power lines can light a lamp continuously at low intensity, depending on the intensity of the electrostatic field.

Also, placing a bulb half way up a two-way radio antenna while transmitting will illuminate the bulb due to the RF energy.

Ion Thruster

An ion thruster is a form of electric propulsion used for spacecraft propulsion that creates thrust by acceleratingions. Ion thrusters are categorised by how they accelerate the ions, using either electrostatic or electromagnetic force. Electrostatic ion thrusters use the Coulomb force and accelerate the ions in the direction of the electric field. Electromagnetic ion thrusters use the Lorentz force to accelerate the ions. The term "ion thruster" by itself usually denotes the electrostatic or gridded ion thrusters.

The thrust created in ion thrusters is very small compared to conventional chemical rockets, but has a very high specific impulse, or propellant efficiency. This high propellant efficiency is achieved by accelerating the propellant to very high speed. They do, however, use a large amount of power, and in use their performance is power-limited (whereas normal chemical thrusters are energy-limited). Given the practical weight of suitable power sources, the accelerations given by these types of thrusters is of the order of one thousandth of standard gravity.

Due to their relatively high power needs, given the specific power of power supplies, and the requirement of an environment void of other ionised particles, ion thrust propulsion is currently only practical in space.

Origins

The official father of the concept of electric propulsion is Konstantin Tsiolkovsky as he is the first to publish mention of the idea in 1911. However, the first documented instance where the possibility of electric propulsion is considered is found in Robert H. Goddard's handwritten notebook in an entry dated 6 September 1906. The first experiments with ion thrusters were carried out by Goddard at Clark University from 1916–1917. The technique was recommended for near-vacuum conditions at high altitude, but thrust was demonstrated with ionised air streams at atmospheric pressure. The idea appeared again in Hermann Oberth's "Wege zur Raumschiffahrt" (Ways to Spaceflight), published in 1923, where he explained his thoughts on the mass savings of electric propulsion, predicted its use in spacecraft propulsion and attitude control, and advocated electrostatic acceleration of charged gases.

A working ion thruster was built by Harold R. Kaufman in 1959 at the NASA Glenn Research Centre facilities. It was similar to the general design of a gridded electrostatic ion thruster with mercury as its fuel. Suborbital tests of the engine followed during the 1960s and in 1964 the engine was sent into a suborbital flight aboard the Space Electric Rocket Test 1 (SERT 1). It successfully operated for the planned 31 minutes before falling back to Earth.

The Hall effect thruster was studied independently in the U.S. and the USSR in the 1950s and 60s. However, the concept of a Hall thruster was only developed into an efficient propulsion device in the former Soviet Union, whereas in the U.S. scientists focused instead on developing gridded ion thrusters. Hall effect thrusters were operated on Soviet satellites since 1972. Until the 1990s they were mainly used for satellite stabilisation in North-South and in East-West directions. Some 100-200 engines completed their mission on Soviet and Russian satellites until the late 1990s. Soviet thruster design was introduced to the West in 1992 after a team of electric propulsion specialists, under the support of the Ballistic Missile Defence Organisation, visited Soviet laboratories.

General Description

Ion thrusters use beams of ions (electrically charged atoms or molecules) to create thrust in accordance with momentum conservation. The method of accelerating the ions varies, but all designs take advantage of the charge/mass ratio of the ions. This ratio means that relatively small potential differences can create very high exhaust velocities. This reduces the amount of reaction mass or fuel required, but increases the amount of specific power required compared to chemical rockets. Ion thrusters are therefore able to achieve extremely high specific impulses. The drawback of the low thrust is low spacecraft acceleration because the mass of current electric power units is directly correlated with the amount of power given. This low thrust makes ion thrusters unsuited for launching spacecraft into orbit, but they are ideal for in-space propulsion applications.

Various ion thrusters have been designed and they all generally fit under two categories. The thrusters are categorised as either electrostatic or electromagnetic. The main difference is how the ions are accelerated.

- Electrostatic ion thrusters use the Coulomb force and are categorised as accelerating the ions in the direction of the electric field.
- Electromagnetic ion thrusters use the Lorentz force to accelerate the ions.

Power supplies for ion thrusters are usually solar panels, but at sufficiently large distances from the Sun, nuclear power is used. In each case the power supply mass is essentially proportional to the peak power that can be supplied, and they both essentially give, for this application, no limit to the energy.

Electrostatic Ion Thrusters

Gridded Electrostatic ion Thrusters

Gridded electrostatic ion thrusters commonly utilise xenon gas. This gas has no charge and is ionised by bombarding it with energetic electrons. These electrons can be provided from a hot cathode filament and accelerated in the electrical field of the cathode fall to the anode (Kaufman type ion thruster). Alternatively, the electrons can be accelerated by the oscillating electric field induced by an alternating magnetic field of a coil, which results in a self-sustaining discharge and omits any cathode (radiofrequency ion thruster).

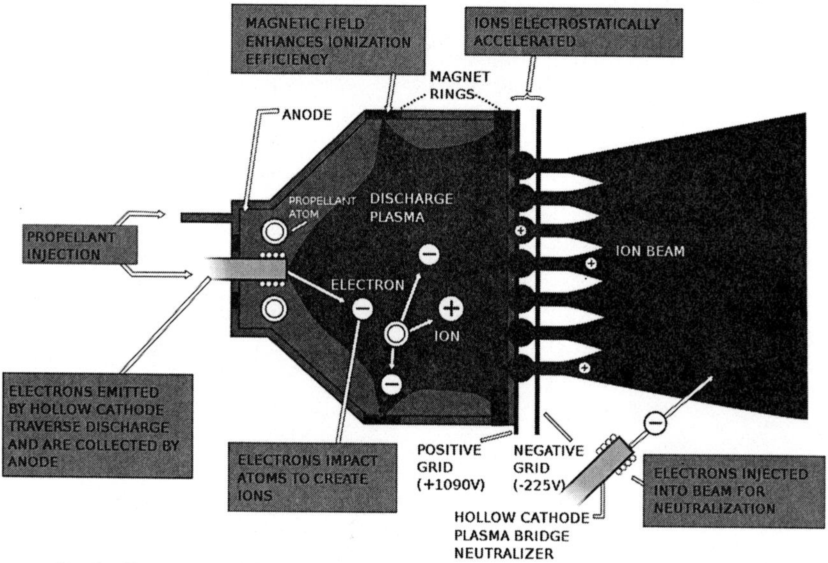

Figure 9: A diagram of how a gridded electrostatic ion engine (Kaufman type) works

The positively charged ions are extracted by an extraction system consisting of 2 or 3 multi-aperture grids. After entering the grid system via the plasma sheath the ions are accelerated due to the potential difference between the first and second grid (named screen and accelerator grid) to the final ion energy of typically 1-2 keV, thereby generating the thrust.

Ion thrusters emit a beam of positive charged xenon ions only. To avoid charging-up the spacecraft, another cathode is placed near the engine, which emits electrons (basically the electron current is the same as the ion current) into the ion beam. This also prevents the beam of ions from returning to the spacecraft and thereby cancelling the thrust.

Gridded electrostatic ion thruster research (past/present):

- NASA Solar electric propulsion Technology Application Readiness (NSTAR)
- NASA's Evolutionary Xenon Thruster (NEXT)
- Nuclear Electric Xenon Ion System (NEXIS)
- High Power Electric Propulsion (HiPEP)

- EADS Radio-Frequency Ion Thruster (RIT)
- Dual-Stage 4-Grid (DS4G)

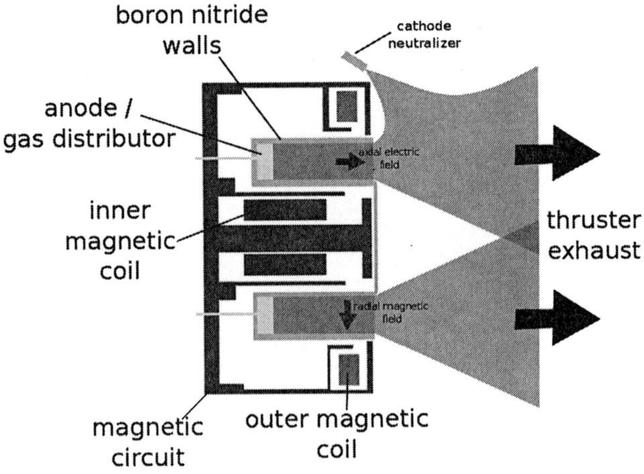

Figure 10: Schematic of a Hall Thruster

Hall Effect Thrusters

Hall effect thrusters accelerate ions with the use of an electric potential maintained between a cylindrical anode and a negatively charged plasma that forms the cathode. The bulk of the propellant (typically xenon gas) is introduced near the anode, where it becomes ionised, and the ions are attracted towards the cathode, they accelerate towards and through it, picking up electrons as they leave to neutralise the beam and leave the thruster at high velocity.

The anode is at one end of a cylindrical tube, and in the centre is a spike that is wound to produce a radial magnetic field between it and the surrounding tube. The ions are largely unaffected by the magnetic field, since they are too massive. However, the electrons produced near the end of the spike to create the cathode are far more affected and are trapped by the magnetic field, and held in place by their attraction to the anode. Some of the electrons spiral down towards the anode, circulating around the spike in a Hall current. When they reach the anode they impact the uncharged propellant and cause it to be ionised, before finally reaching the anode and closing the circuit.

Field Emission Electric Propulsion

Field emission electric propulsion (FEEP) thrusters use a very simple system of accelerating liquid metal ions to create thrust. Most designs use either caesium or indium as the propellant. The design comprises a small propellant reservoir that stores the liquid metal, a narrow tube that the liquid flows through, and an accelerator ring a millimetre past the tube end. Caesium and indium are used due to their high atomic weights, low ionisation potentials, and low melting points. Once the liquid metal reaches the end of the tube, an electric field applied between the emitter and the accelerator ring causes the liquid surface to deform and emit positive ions. The electric field created by the emitter and the accelerator ring then accelerate the ions. An external source of electrons then neutralises the positively charged ion stream to prevent charging of the spacecraft hull.

Electromagnetic Thrusters

Pulsed Inductive Thrusters (PIT)

Pulsed inductive thrusters (PIT) use pulses of thrust instead of one continuous thrust, and have the ability to run on power levels in the order of Megawatts (MW). PITs consist of a large coil encircling a cone shaped tube that emits the propellant gas. Ammonia is the gas commonly used in PIT engines. For each pulse of thrust the PIT gives, a large charge first builds up in a group of capacitors behind the coil and is then released. This creates a current that moves circularly in the direction of jθ. The current then creates a magnetic field in the outward radial direction (Br), which then creates a current in the ammonia gas that has just been released in the opposite direction of the original current. This opposite current ionises the ammonia and these positively charged ions are accelerated away from the PIT engine due to the electric field jθ crossing with the magnetic field Br, which is due to the Lorentz Force.

Magnetoplasmadynamic (MPD) / Lithium Lorentz Force Accelerator (LiLFA)

Magnetoplasmadynamic (MPD) thrusters and lithium Lorentz force accelerator (LiLFA) thrusters use roughly the same idea with the LiLFA thruster building off of the MPD thruster. Hydrogen, argon, ammonia, and nitrogen gas can be used as propellant. In a certain configuration, the ambient gas in Low Earth Orbit (LEO) can be used as a propellant. The gas first enters the main chamber where it is ionised into plasma by the electric field between the anode and the cathode. This plasma then conducts electricity between the anode and the cathode. This new current creates a magnetic field around the cathode, which crosses with the electric field, thereby accelerating the plasma due to the Lorentz Force. The LiLFA thruster uses the same general idea as the MPD thruster, except for two main differences. The first difference is that the LiLFA uses lithium vapour, which has the advantage of being able to be stored as a solid. The other difference is that the cathode is replaced by multiple smaller cathode rods packed into a hollow cathode tube. The cathode in the MPD thruster is easily corroded due to constant contact with the plasma. In the LiLFA thruster the lithium vapour is injected into the hollow cathode and is not ionised to its plasma form/corrode the cathode rods until it exits the tube. The plasma is then accelerated using the same Lorentz Force.

Electrodeless Plasma Thrusters

Electrodeless plasma thrusters have two unique features: the removal of the anode and cathode electrodes and the ability to throttle the engine. The removal of the electrodes takes away the factor of erosion, which limits lifetime on other ion engines. Neutral gas is first ionised by electromagnetic waves and then transferred to another chamber where it is accelerated by an oscillating electric and magnetic field, also known as the ponderomotive force. This separation of the ionisation and acceleration stage give the engine the ability to throttle the speed of propellant flow, which then changes the thrust magnitude and specific impulse values.

Electrothermal Thrusters

Electrothermal thrusters use electric power to accelerate propellant. There are several types:

- Resistojet
- Arcjet
- Microwave electrothermal thrusters
- Ion Cyclotron Heating thrusters (VASIMR)

Helicon Double Layer Thruster

A helicon double layer thruster is a type of plasma thruster, which ejects high velocity ionised gas to provide thrust to a spacecraft. In this thruster design, gas is injected into a tubular chamber (the source *tube*) with one open end. Radio frequency AC power (at 13.56 MHz in the prototype design) is coupled into a specially shaped antenna wrapped around the chamber. The electromagnetic wave emitted by the antenna causes the gas to break down and form a plasma. The antenna then excites a Helicon wave in the plasma, which further heats the plasma. The device has a roughly constant magnetic field in the source tube (supplied by Solenoids in the prototype), but the magnetic field diverges and rapidly decreases in magnitude away from the source region, and might be thought of as a kind of magnetic nozzle. In operation, there is a sharp boundary between the high density plasma inside the source region, and the low density plasma in the exhaust, which is associated with a sharp change in electrical potential. The plasma properties change rapidly across this boundary, which is known as a *current free electric double layer*. The electrical potential is much higher inside the source region than in the exhaust, and this serves both to confine most of the electrons, and to accelerate the ions away from the source region. Enough electrons escape the source region to ensure that the plasma in the exhaust is neutral overall.

Comparisons

The following table compares actual test data of some ion thrusters:

Engine	*Propellant*	*Required Power(kW)*	*Specific Impulse(s)*	*Thrust (mN)*
NSTAR	Xenon	2.3	3,300	92 max.
NEXT	Xenon	7.7	4,300	236 max.
NEXIS	Xenon	20.5	6,000-7,500	400
HiPEP	Xenon	25-50	6,000-9,000	460-670
RIT 22	Xenon	5	3,000-6,000	50 - 200
Hall effect	Bismuth	25	3,000	1,130
Hall effect	Bismuth	140	8,000	2,500
Hall effect	Xenon	25	3,250	950
Hall effect	Xenon	75	2,900	2,900
FEEP	Liquid Caesium	$6x10^{-5}$-0.06	6,000-10,000	0.001-1
VASIMR	Argon	200	3,000-30,000	~5000

The following thrusters are highly experimental and have been tested only in pulse mode.

Engine	*Propellant*	*Required Power(kW)*	*Specific Impulse(s)*	*Thrust (mN)*
MPDT	Hydrogen	1,500	4,900	26,300
MPDT	Hydrogen	3,750	3,500	88,500
MPDT	Hydrogen	7,500	6,000	60,000
LiLFA	Lithium Vapour	500	4,077	12,000

Lifetime

A major limiting factor of ion thrusters is their small thrust; however, it is generated at a high propellant efficiency (mass utilisation, specific impulse). The efficiency comes from the high exhaust velocity, which in turn demands high energy, and the performance is ultimately limited by the available spacecraft power.

The low thrust requires ion thrusters to provide continuous thrust for a long time to achieve the needed change in velocity (delta-v) for a particular mission. To cause enough change in momentum, ion thrusters are designed to last for periods of weeks to years.

In practice the lifetime of electrostatic ion thrusters is limited by several processes:

- In electrostatic gridded ion thruster design, charge-exchange ions produced by the beam ions with the neutral gas flow can be accelerated towards the negatively biased accelerator grid and cause grid erosion. End-of-life is reached when either a structural failure of the grid occurs or the holes in the accelerator grid become so large that the ion extraction is largely affected (e.g. by the occurrence of electron backstreaming). Grid erosion cannot be avoided and is the major lifetime-limiting factor. By a thorough grid design and material selection, lifetimes of 20,000 hours and far beyond are reached, which is sufficient to fulfill current space missions.

A test of the NASA Solar electric propulsion Technology Application Readiness (NSTAR) electrostatic ion thruster resulted in 30,472 hours (roughly 3.5 years) of continuous thrust at maximum power. The test was concluded prior to any failure and test results showed the engine was not approaching failure either.

- Hall thrusters suffer from very strong erosion of the ceramic discharge chamber. Due to the rather high discharge voltages of up to 1000V energetic ions can impinge to the chamber walls and erode material. Lifetimes of a few thousand hours are reached.

Propellants

Ionisation energy represents a very large percentage of the energy needed to run ion drives. The ideal propellant for ion drives is thus a propellant molecule or atom that is easy to ionise, that has a high mass/ionisation energy ratio. In addition, the propellant should not cause erosion of the thruster to any great degree to permit long life; and should not contaminate the vehicle.

Many current designs use xenon gas as it is easy to ionise, has a reasonably high atomic number, its inert nature, and low erosion. However, xenon is globally in short supply and very expensive.

Older designs used mercury, but this is toxic and expensive, tended to contaminate the vehicle with the metal and was difficult to feed accurately.

Other propellants such as bismuth show promise and are areas of research, particularly for gridless designs such as Hall effect thrusters.

VASIMR design (and other plasma based engines) are theoretically able to use practically any material for propellant. However, in current tests the most practical propellant is argon, which is a relatively abundant and cheap gas.

Energy Efficiency

Ion thrusters are frequently quoted with an efficiency metric. This efficiency is the kinetic energy of the exhaust jet emitted per second divided by the electrical power into the device.

The actual overall system energy efficiency in use is determined by the propulsive efficiency, which depends on vehicle speed and exhaust speed. Some thrusters can vary exhaust speed in operation, but all can be designed with different exhaust speeds. At the lower end of Isps the overall efficiency drops because the ionisation takes up a larger percentage energy, and at the high end propulsive efficiency is reduced.

Optimal efficiencies and exhaust velocities can thus be calculated for any given mission to give minimum overall cost.

Applications

Ion thrusters have many applications for in-space propulsion. The best applications of the thrusters make use of the long lifetime when significant thrust is not needed. Examples of this include orbit transfers, attitude adjustments, drag compensation for low earth orbits, transporting cargo such as chemical fuels between propellant depots and ultra fine adjustments for more scientific missions. Ion thrusters can also be used for interplanetary and deep space missions where time is not crucial. Continuous thrust over a very long time can build up a larger velocity than traditional chemical rockets.

Missions

Of all the electric thrusters, ion thrusters have been the most seriously considered commercially and academically in the quest for interplanetary missions and orbit raising maneuvers. Ion thrusters are seen as the best solution for these missions as they require very high change in velocity overall that can be built up over long periods of time.

Operational Missions

Several spacecraft have operated with this technology.

SERT

Ion propulsion systems were first demonstrated in space by the NASA Lewis (now Glenn Research Centre) missions "Space Electric Rocket Test" (SERT) I and II. The first was SERT-1, launched July 20, 1964, successfully proved that the technology operated as predicted in space. These were electrostatic ion thrusters using mercury and cesium as the reaction mass. The second test, SERT-II, launched on February 3, 1970, verified the operation of two mercury ion engines for thousands of running hours.

Deep Space 1

NASA has developed an ion thruster called NSTAR for use in their interplanetary science missions beginning in the late-1990s. This xenon-propelled ion thruster was first space-tested in the highly successful space probe Deep Space 1, launched in 1998. This was the first use of electric propulsion as the interplanetary propulsion system on a science mission. Based on the NASA design criteria, Hughes Research Labs, developed the XIPS (Xenon Ion Propulsion System) for performing station keeping on geosynchronous satellites.. Hughes (EDD) manufactured the NSTAR thruster used on the spacecraft.

Artemis

On 12 July 2001, the European Space Agency failed to launch their Artemis telecommunication satellite to desired altitude, and left it in a decaying orbit. The satellite's chemical propellant supply was sufficient to transfer it to a semi-stable orbit, and over the next 18 months the experimental on board ion propulsion system RIT-10 (intended only for secondary station keeping and maneuvering) was utilised to transfer it to a geostationary orbit.

Hayabusa

The Japanese space agency's Hayabusa, which was launched in 2003 and successfully rendezvoused with the asteroid 25143

Itokawa and remained in close proximity for many months to collect samples and information, was powered by four xenon ion engines. It used xenon ions generated by microwave electron cyclotron resonance, and a carbon / carbon-composite material (which is resistant to erosion) for its acceleration grid. Although the ion engines on Hayabusa had some technical difficulties, in-flight reconfiguration allowed one of the four engines to be repaired, and allowed the mission to successfully return to Earth.

Smart 1

The Hall effect thruster is a type of ion thruster that has been used for decades for station keeping by the Soviet Union and is now also applied in the West: the European Space Agency's satellite Smart 1, launched in 2003, used it (Snecma PPS-1350-G). This satellite completed its mission on 3 September 2006, in a controlled collision on the Moon's surface, after a trajectory deviation so scientist could see the 3 meter crater the impact created on the visible side of the moon.

Dawn

Dawn was launched on 27 September 2007 to explore the asteroid Vesta and the dwarf planet Ceres. To cruise from Earth to its targets it uses three Deep Space 1 heritage xenon ion thrusters (firing only one at a time) to take it in a long outward spiral. An extended mission in which Dawn explores other asteroids after Ceres is also possible. Dawn's ion drive is capable of accelerating from 0 to 60 mph (97 km/h) in 4 days.

GOCE

ESA's Gravity Field and Steady-State Ocean Circulation Explorer was launched on March 16, 2009. It will continue to use ion propulsion throughout its twenty month mission to combat the air-drag it experiences in its low orbit.

Planned Missions

In addition, several missions are planned to use ion thrusters in the next few years.

BepiColombo

ESA will launch the BepiColombo mission to Mercury in 2014. It uses ion thrusters in combination with swing-bys to get to Mercury, where a chemical rocket will be fired for orbit insertion.

GSAT-4

Indian Space Research Organisation will utilise ion thrusters in its GSAT-4 satellite. This will increase the life of satellite from the present 10 years to 15 years.

LISA Pathfinder

LISA Pathfinder is an ESA spacecraft to be launched in 2013. It will not use ion thrusters as its primary propulsion system, but will use both colloid thrusters and FEEP for very precise altitude control—the low thrusts of these propulsion devices make it possible to move the spacecraft incremental distances very accurately. It is a test for the possible LISA mission.

International Space Station

As of February 2010, a 2011/2012 launch of an Ad Astra VF-200 200 kW VASIMR electromagnetic thruster is planned for placement and testing on the International Space Station. The VF-200 is a flight version of the VX-200. though it may be later. Since the available power from the ISS is less than 200 kW, the ISS VASIMR will include a trickle-charged battery system allowing for 15 min pulses of thrust. Testing of the engine on ISS is valuable because ISS orbits at a relatively low altitude and experiences fairly high levels of atmospheric drag, making periodic boosts of altitude necessary. Currently, altitude reboosting by chemical rockets fulfills this requirement. If the tests of VASIMR reboosting of the ISS goes according to plan, the increase in specific impulse could mean that the cost of fuel for altitude reboosting will be one-twentieth of the current $210 million annual cost. Hydrogen is generated by the ISS as a by-product, which is currently vented into space.

Neon Sign

Neon signs are made using electrified, luminous tube lights that contain rarefied neon or other gases. They are the most common use for neon lighting, which was first demonstrated in a modern form in December, 1910 by Georges Claude at the Paris Motor Show. While they are used worldwide, neon signs were extremely popular in the United States from about 1920–1960. The installations in Times Square were famed, and there were nearly 2000 small shops producing neon signs by 1940. In addition to signage, neon lighting is now used frequently by artists and architects, and (in a modified form) in plasma display panels and televisions. The signage industry has declined in the past several decades, and cities are now concerned with preserving and restoring their antique neon signs.

History

The neon sign is an evolution of the earlier Geissler tube, which is an electrified glass tube containing a "rarefied" gas (the gas pressure in the tube is well below atmospheric pressure). When a voltage is applied to electrodes inserted through the glass, an electrical glow discharge results. Geissler tubes were quite popular in the late 1800s, and the different colours they emitted were characteristics of the gases within. They were, however, unsuitable for general lighting; the pressure of the gas inside typically declined in use. The direct predecessor of neon tube lighting was the Moore tube, which used nitrogen or carbon dioxide as the luminous gas and a patented mechanism for maintaining pressure; Moore tubes were sold for commercial lighting for a number of years in the early 1900s.

The discovery of neon in 1898 included the observation of a brilliant red glow in Geissler tubes. Immediately following neon's discovery, neon tubes were used as scientific instruments and novelties. A sign created by Perley G. Nutting and displaying the word "neon" may have been shown at the Louisiana Purchase Exposition of 1904, although this claim has been disputed; in any event, the scarcity of neon would have precluded the development of a lighting product. However, after 1902, Georges Claude's company in France, Air Liquide, began producing industrial quantities of neon, essentially as a byproduct of their air liquefaction business. From December 3–18, 1910, Claude demonstrated two 12-metre (39 ft) long bright red neon tubes at the Paris Motor Show. This demonstration lit a peristyle of the *Grand Palais* (a large exhibition hall). Claude's associate, Jacques Fonseque, realised the possibilities for a business based on signage and advertising. By 1913 a large sign for the vermouth Cinzano illuminated the night sky in Paris, and by 1919 the entrance to the Paris Opera was adorned with neon tube lighting. Over the next several years, patents were granted to Claude for two innovations still used today: a "bombardment" technique to remove impurities from the working gas of a sealed sign, and a design for the internal electrodes of the sign that prevented their degradation by sputtering.

In 1923, Georges Claude and his French company Claude Neon introduced neon gas signs to the United States by selling two to a Packard car dealership in Los Angeles. Earle C. Anthony purchased the two signs reading "Packard" for $1,250 apiece. Neon lighting quickly became a popular fixture in outdoor advertising. Visible even in daylight, people would stop and stare at the first neon signs for hours, dubbed "liquid fire."

The next major technological innovation in neon lighting and signs was the development of fluorescent tube coatings. Jacques Risler received a French patent in 1926 for these. Neon signs that use an argon/mercury gas mixture emit a good deal of ultraviolet light. When this light is absorbed by a fluorescent coating, preferably inside the tube, the coating (called a "phosphor") glows with its own colour. While only a few colours were initially available to sign designers, after the Second World War (1939–1945) phosphor materials were researched intensively for use in colour televisions. About two dozen colours were available to neon sign designers in the 1960s, and today there are nearly 100 available colours.

Fabrication

Neon tube signs are produced by the craft of bending glass tubing into shapes. A worker skilled in this craft is known as a glass bender, neon bender or tube bender. The neon tube is made out of 4-5' straight sticks of hollow glass sold by sign suppliers to neon shops worldwide, where they are manually assembled into individual custom designed and fabricated lamps. There are many dozens of colours available, determined by the type of glass tubing and the composition of the gas filling.

Neon sign manufacturing is a cottage industry and an eclectic art, and in most cases, is organised as a small family business. Even today, almost all neon tubes are hand made and labour intensive. The shop equipment used to fabricate signs is itself typically custom assembled from scratch from discrete parts by the craftsmen who will use the equipment.

Tubing in external diameters ranging from about 8–15 mm with a 1 mm wall thickness is most commonly used, although 6mm tubing is now commercially available in coloured glass tubes. The tube is heated in sections using several types of burners that are selected according to the amount of glass to be heated for each bend. These burners include ribbon, cannon, or crossfires, as well as a variety of gas torches. Ribbon burners are strips of fire that make the gradual bends while crossfires, when used, make the sharp bends.

The interior of the tubes may be coated with a thin phosphorescent powder coating, affixed to the interior wall of the tube by a binding material. The tube is filled with a purified gas mixture, and the gas ionised by a high voltage applied between the ends of the sealed tube through cold cathodes welded onto the ends. The colour of the light emitted by the tube may be just that coming from the gas, or the light from the phosphor layer. Different phosphor-coated tubing sections may be butt welded together using glass working torches to form a single tube of varying colours, for effects such as a sign where each letter displays a different colour letter within a single word, such as shown in the sign in the photo above right.

"Neon" is used to denote the general type of lamp, but neon gas is only one of the types of tube gases principally used in commercial application. Pure neon gas is used to produce only about a third of the colours. The greatest number of colours is produced by filling with another inert gas, argon, and a drop of mercury (Hg) which is added to the tube immediately after purification. When the tube is ionised by electrification, the mercury evaporates into mercury vapour, which fills the tube and produces strong ultraviolet light. The ultraviolet light thus produced excites the various phosphor coatings designed to produce different colours. Even though this class of neon tubes use no neon at all, they are still denoted as "neon." Mercury-bearing lamps are a type of cold-cathode fluorescent lamps.

Each type of neon tubing produces two completely different possible colours, one with neon gas and the other with argon/mercury. Some "neon" tubes are made without phosphor coatings for some of the colours. Clear tubing filled with neon gas produces the ubiquitous yellowish orange colour with the interior plasma column clearly visible, and is the cheapest and simplest tube to make. Traditional neon glasses in America over 20 years old are lead glass that are easy to soften in gas fires, but recent environmental and health concerns of the workers has prompted manufacturers to seek more environmentally safe special soft glass formulas. One of the vexing problems avoided this way is lead glass' tendency to burn into a black spot emitting lead fumes in a bending flame too rich in the fuel/oxygen mixture. Another traditional line of glasses was coloured soda lime glasses coming in a myriad of glass colour choices, which produce the highest quality, most hypnotically vibrant and saturated hues. Still more colour choices are afforded in either coating, or not coating, these coloured glasses with the various available exotic phosphors.

Long Lifetime

It is the wide range of colours and the ability to make a tube that can last for years if not decades without replacement, that makes this an art. Since these tubes require so much custom labour, they would have very little economic viability if they did not have such a long lifetime when well processed. The intensity of neon light produced increases slowly as the tube diameter grows smaller, that is, the intensity varies inversely with the square root of the interior diameter of the tubing, and the resistance of the tube increases as the tubing diameter decreases accordingly, because tube ionisation is greatest at the centre of the tube, and the ions migrate to and are recaptured and neutralised at the tube walls. The greatest cause of neon tube failure is the gradual absorption of neon gas by high voltage ion implantation into the interior glass walls of the tubes which depletes the gas, and eventually causes the tube resistance to rise to a level that it can no longer light at the rated voltage, but this may take 7–10 years.

The actual cause of 80% of neon sign failures is the burnout of the high voltage electrical wires connecting the tubes inside of metal conduits. A very common type of neon sign is made from a formed metal box having a coloured translucent face, called "channel lettering". Newer channel letter signs are being replaced by high brightness LEDs.

This long lifetime has created a practical market for neon use for interior architectural cove lighting in a wide variety of uses including homes, where the tube can be bent to any shape, fitted in a small space, and can do so without requiring tube replacement for a decade or more.

Tube Bending

A section of the glass is heated until it is malleable; then it is bent into shape and aligned to a pattern containing the graphics or lettering that the final product will ultimately conform to. This is where the real art of neon comes in that takes some artisans from a year up to several years of practice to master. A tube bender corks off the hollow tube before heating and holds a latex rubber blow hose at the other end, through which he gently presses a small amount of air in order to keep the tube diameter constant as it is bending. The trick of bending is to bend one small section or bend at a time, heating one part of the tubing so that it is soft, without heating some other part of the tube as well, which would make the bend uncontrollable. A bend, once the glass is heated, must be brought to the pattern and fitted rapidly before the glass hardens again because it is difficult to reheat once completely cooled without risking breakage. It is frequently necessary to skip one or more bends and come back to it later, by measuring carefully along the length of the tube. One tube letter may contain 7-10 small bends, and mistakes are not easily corrected without going back and starting all over again. If more tubing is required, another piece is welded onto it, or the parts can be all welded onto each other at the final step. The finished tube must be absolutely vacuum tight in order to operate, and it must be vacuum clean inside. Once the tube is filled with mercury, if any mistake is made after that, the entire tube had, or should, be started over again, because breathing heated mercury impregnated glass and phosphor causes long term heavy metal poisoning in neon workers. Sticks of tubing are joined until the tube reaches an impractical size, and several tubes are joined in series with the high voltage neon transformer. Extreme ends of the electrical circuit must be isolated from each other to prevent tube puncture and buzzing from corona effect.

Bombardment

A cold cathode electrode is melted (or welded) to each end of the tube as it is finished. The electrodes are also traditionally lead glass and contain a small metal shell with two wires protruding through the glass to which the sign wiring will later be attached. All welds and seals must be perfectly leak-proof to high vacuum before proceeding further.

The tube is attached to a manifold which is itself attached to a high-quality vacuum pump. The tube is then evacuated of air until it reaches near-vacuum. During evacuation, a high current is forced through the tube via the wires protruding from each electrode (in a process known as "bombarding"). This current and voltage is far above the level that occurs in final operation of the tube. The current depends on the specific electrodes used and the diameter of the tube, but is typically in the 450 mA to 800 mA range, at an applied voltage usually between 22,000-26,000 V. The bombarding transformer acts as an adjustable constant current source, and the voltage produced depends on the length and pressure of the tube. Typically the operator will maintain the pressure as high as the bombarder will allow to ensure maximum power dissipation and heating.

This very high power dissipation in the tube heats the glass walls to a temperature of several hundred degrees Celsius, and any dirt and impurities within are drawn off in the gasified form by the vacuum pump. The greatest impurities that are driven off this way are the gases that coat the inside wall of the tubing by adsorption, mainly oxygen, carbon dioxide, and especially water vapour. The current also heats the electrode metal to over 600°C, producing a bright orange incandescent colour. The cathodes are prefabricated hollow metal shells with a small opening (sometimes a ceramic donut aperture) which contains in the interior surface of the shell a light dusting of a cold cathode low work function powder (usually a powder ceramic molar eutectic point mixture including $BaCO_2$, combined with other alkaline earth oxides, which reduces to BaO_2 when heated to about 500 degrees F, and reduces the work function of the electrode for cathodic emission. Barium Oxide has a work function of roughly 2 whereas tungsten at room temperature has a work function exponentially 100 times more, or 4.0. This represents the cathode drop or electron energy required to remove electrons from the surface of the cathode. This avoids the necessity of using a hot wire thermoelectric cathode such as is used in conventional fluorescent lamps. And for that reason, neon tubes are extremely long lived when properly processed, in contrast to fluorescent tubing, because there is no wire filament as there is in a fluorsecent tube to burn out like a common light bulb. The principal purpose of doing this is to purify the interior of the tube *before* the tube is sealed off so that when it is operated, these gases and impurities are not driven off and released by the plasma and the heat generated into the sealed tube, which would quickly burn the metal cathodes and mercury droplets (if pumped with argon/mercury) and oxidise the interior gases and cause immediate tube failure. The more thorough the purification of the tube is, the longer lasting and stable the tube will be in actual operation. Once these gases and impurities are liberated under pre-filling bombardment into the tube interior they are quickly evacuated by the pump.

While still attached to the manifold, the tube is allowed to cool while pumping down to the lowest pressure the system can achieve. It is then filled to a low pressure of a few torrs (millimeters of mercury) with one of the noble gases, or a mixture of them, and sometimes a small amount of mercury. This gas fill pressure represents roughly 1/100th of the pressure of the atmosphere. The required pressure depends on the gas used and the diameter of the tube, with optimal values ranging from 6 Torr (0.8 kPa) (for a long 20 mm tube filled with argon/mercury) to 27 Torr (3.6 kPa) (for a short 8 mm diameter tube filled with pure neon). Neon or argon are the most common gases used; krypton, xenon, and helium are used by artists for special purposes but are not used alone in normal signs. A premixed combination of argon and helium is often used in lieu of pure argon when a tube is to be installed in a cold climate, since the helium increases voltage drop (and thus power dissipation), warming the tube to operating temperature faster. Neon glows bright red or reddish orange when lit. When argon or argon/helium is used, a tiny droplet of mercury is added. Argon by itself is very dim pale lavender when lit, but the droplet of mercury fills the tube with mercury vapour when sealed, which then emits ultraviolet light upon electrification. This ultraviolet emission allows finished argon/mercury tubes to glow with a variety of bright colours when the tube has been coated on the interior with ultraviolet-sensitive phosphors after being bent into shape.

Heat Processed Neon Tubes

An alternate way of processing finished neon tubes has also been used. Because the only purpose of bombardment by electrical means is to purify the interior of tubes, it is also possible to produce a tube by heating the tube externally either with a torch or with an oven, while heating the electrode with a Radio Frequency Induction Heating coil (RFIH). While this is less productive, it creates a cleaner custom tube with significantly less cathode damage, longer life and brilliance, and can produce tubes of very small sizes and diameters, down to 6mm OD. The tube is heated thoroughly under high vacuum without external electrical application, until the outgassed gases can be seen to have been totally depleted and the pressure drops to a high vacuum again. Then the tube is filled, sealed and the mercury dropped and shaken.

Electrical Wiring

The finished glass pieces are illuminated by either a neon sign transformer or a switching power supply running at voltages ranging between 3-15 kV and currents between 20 and 120 mA. These power supplies operate as constant-current sources (a high voltage supply with a very high internal impedance), since the tube has a negative characteristic electrical impedance. Standard tube tables established in the early days of neon are still used that specify the gas fill pressures, in either Ne or Hg/Ar, as a function of tube length in feet, tube diameter and transformer voltage.

The standard traditional neon transformer, a magnetic shunt transformer, is a special non-linear type designed to keep the voltage across the tube raised to whatever level is necessary to produce the fixed current needed, up to the maximum limit of the neon footage possible.

Newer, compact high frequency inverter-converter transformers developed in the early '90s are used as well, especially when low Radio Frequency Interference (RFI) is needed, such as in locations near high-fidelity sound equipment. The reason for this is that at 20 kHz, the typical frequency of these solid state transformers, the plasma electron-ion recombination time is too short to extinguish and reignite the plasma in the tube at every 1/120th second as in 50/60 Hz transformers and so the plasma does not broadcast high frequency switching noise. The plasma simply remains lit at all times, becoming radio noise free. The most common current rating is 30 mA for general use, with 60 mA used for high-brightness applications like channel letters or architectural lighting. 120 mA sources are occasionally seen in illuminating applications, but are uncommon since special electrodes are required to withstand the current, and an accidental shock from a 120 mA transformer is much more likely to be fatal than from the lower current supplies.

The efficiency of neon lighting ranges between that of ordinary incandescent lights and that offluorescent lamps, depending on colour. On a per-watt basis, incandescents produce 10 to 20 lumens, while fluorescents produce 50 to 100 lumens. Neon light efficiency ranges from 10 lumens per watt for red, up to 60 lumens for green and blue when these colours result from internal phosphor coatings.

Blocking out and Coating

A trick of the eye is used to produce visually distinct neon display segments by blocking out parts of the tube with an opaque coating. One complete assembly may be composed of contiguous tube elements joined by glass welding to one another so that the same current passes through, for example, several letters joined end to end from cathode to cathode. To the untrained eye, this looks like separate tubes, but the electrical splice is the plasma inside the crossover glass itself. The entire tube lights up, but the segments that the viewer is not supposed to see are covered with highly opaque special black or gray glass paint. This heat-resistant coating is either painted on or dipped. Without blockout paint, the unintended visual connections would make the display appear confusing.

In most mass produced low-priced signs today, clear glass tubing is coated with translucent paint to produce coloured light. In this way, several different colours can be produced inexpensively from a single glowing tube. Over time, elevated temperatures, thermal cycling, or exposure to weather may cause the coloured coating to flake off the glass or change its hue. A more expensive alternative is to use high-quality coloured glass tubing, which retains a more stable appearance as it ages.

Applications

The light-emitting tubes form coloured lines with which a text can be written or a picture drawn, including various decorations, especially in advertising and commercial signage. By programming sequences of switching parts on and off, there are many possibilities for dynamic light patterns that form animated images.

Neon signs are increasingly replaced with LED signs due to the lower cost and longer operating life of LEDs. This trend seems likely to increase given the steady advance in LED luminosity and decreasing cost of high-intensity LEDs.

Heat Shield

A heat shield is designed to shield a substance from absorbing excessive heat from an outside source by either dissipating, reflecting or simply absorbing the heat. It is often used as a form of Exhaust Heat Management.

Uses

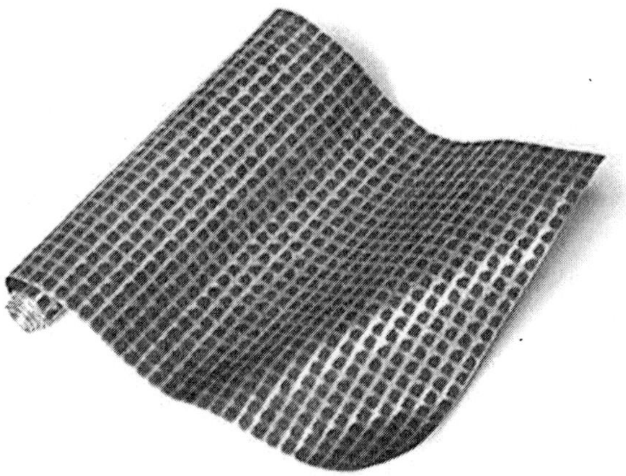

Figure 11: First ever flexible ceramic heat shield

Automotive

Due to the large amounts of heat given off by internal combustion engines, heat shields are used on most engines to protect components and bodywork from heat damage. As well as protection, effective heat shields can give a performance benefit by reducing the underbonnet temperatures, therefore reducing the intake temperature. Heat shields vary widely in price, but most are easy to fit, usually by stainless steel clips or high temperature tape. There are two main types of automotive heat shield:

- The Rigid heat shield has, until recently, been made from solid steel, but is now often made from aluminium. Some high-end rigid heat shields are made out of aluminium sheet or other composites, with a ceramic thermal barrier coating to improve the heat insulation.
- The Flexible heat shield is normally made from thin aluminium sheeting, sold either flat or in a roll, and is bent by hand, by the fitter. High performance flexible heat shields sometimes include extras, such as ceramic insulation applied via plasma spraying. These latest products are commonplace in top-end motorsports such as Formula 1.

As a result, a heat shield is often fitted by both amateur and professional personnel during a phase of engine tuning.

Space

Spacecraft which land on a body with an atmosphere, such as Earth, Mars and Venus, currently do so by entering at high speeds, depending on friction rather than rocket power to slow the craft. A side effect of this is aerodynamic heating, which in the case of an Earth reentry can be destructive. An aerodynamic heat shield consists of a protective layer of special materials to dissipate the heat. Two basic types of aerodynamic heat shield have been used:

- An ablative heat shield consists of a layer of plastic resin, the outer surface of which is heated to a gas which carries the heat away by convection. Such shields were used on the Mercury, Gemini and Apollo spacecraft.
- A thermal soak heat shield uses an insulating material to absorb and radiate the heat away from the spacecraft structure. This type was used on the Space Shuttle, consisting of fibreglass tiles over most of the vehicle surface, with reinforced carbon-carbon material on the highest heat load points, the nose and wing leading edges. Damage to this material on a wing caused the 2003 Space Shuttle Columbia disaster.

Some spacecraft also use a heat shield (in the conventional automotive sense) to protect fuel tanks and equipment from heat produced by a large rocket engine. Such shields were used on the Apollo Service Module and Lunar Module descent stage.

Industry

Heat shields are often used to protect heat-sensitive components used in various industrial processes.

Firearms

Heat shields are often affixed to semi-automatic or automatic rifles and shotguns in order to protect the user's hands from the heat caused by firing shots in rapid succession.

Cold Plasma

In the cold plasma method, pure oxygen gas is exposed to a plasma created by dielectric barrier discharge. The diatomic oxygen is split into single atoms, which then recombine in triplets to form ozone.

Cold plasma machines utilise pure oxygen as the input source and produce a maximum concentration of about 5% ozone. They produce far greater quantities of ozone in a given space of time compared to ultraviolet production. However, because cold plasma ozone generators are very expensive, they are found less frequently than the previous two types.

The discharges manifest as filamentary transfer of electrons (micro discharges) in a gap between two electrodes. In order to evenly distribute the micro discharges, a dielectric insulator must be used to separate the metallic electrodes and to prevent arcing.

Some cold plasma units also have the capability of producing short-lived allotropes of oxygen which include O_4, O_5, O_6, O_7, etc. These species are even more reactive than ordinary O_3.

Fusion Power

Fusion power is the power generated by nuclear fusion processes. In fusion reactions two light atomic nuclei fuse together to form a heavier nucleus (in contrast with fission power). In doing so they release a comparatively large amount of energy arising from the binding energy due to the strong nuclear force which is manifested as an increase in temperature of the reactants. Fusion power is a primary area of research in plasma physics.

The term is commonly used to refer to potential commercial production of net usable power from a fusion source, similar to the usage of the term "steam power." The leading designs for controlled fusion research use magnetic (tokamak design) or inertial (laser) confinement of a plasma, with heat from the fusion reactions used to operate a steam turbine which in turn drives electrical generators, similar to the process used in fossil fuel and nuclear fission power stations.

Fusion power is believed to have significant safety advantages over current power stations based on nuclear fission. Fusion only takes place under very limited and controlled circumstances, and requires a constant feed of new fuel to maintain the reaction, so the cessation of active fuelling or simple changes to the control system quickly shuts down fusion power reactions. By comparison, fission reactors only require that there is sufficient fuel within a small enough space, and are subject to catastrophic failures that self-maintain the reaction, notably a meltdown. In a fusion reactor there is no possibility of runaway heat build-up or large-scale release of radioactivity, little or no atmospheric pollution, the power source comprises light elements in small quantities, the waste products are short-lived in terms of radioactivity.

Fusion powered electricity generation was initially believed to be readily achievable, as fission power had been. However the extreme requirements for continuous reactions and plasma containment led to projections being extended by several decades. In 2010, more than 60 years after the first attempts, commercial power production is still believed to be unlikely before 2050.

As of July 2010, the largest experiment by means of magnetic confinement has been the Joint European Torus (JET). In 1997, JET produced a peak of 16.1 megawatts (21,600 hp) of fusion power (65% of input power), with fusion power of over 10 MW (13,000 hp) sustained for over 0.5 sec. In June 2005, its successor, ITER, was announced by the seven parties involved in the project - the United States, China, the European Union (EU), India, Japan, the Russian Federation, and South Korea. ITER is designed to produce ten times more fusion power than the power put into the plasma over many minutes; for example 50 MW of input power to produce 500 MW of output power. ITER is currently under construction in Cadarache, France. DEMO is intended as the next generation of research from ITER, and to be the first reactor demonstrating sustained net energy-producing fusion on a commercial scale. It has been proposed to begin construction of DEMO in 2024.

Inertial (laser) confinement, which was for a time seen as more difficult or infeasible, has generally seen less development effort than magnetic approaches. However this approach made a comeback following further innovations, and is being developed at both the United States National Ignition Facility as well as the planned European Union High Power laser Energy Research (HiPER) facility. As of 2010 heating to 3.3 million Kelvin was achieved and in October 2010 the first integrated ignition test was announced to have been completed successfully with the 192-beam laser system firing over a million joules of ultraviolet laser energy into a capsule filled with hydrogen fuel. Fusion ignition tests are to follow.

Fuel Cycle

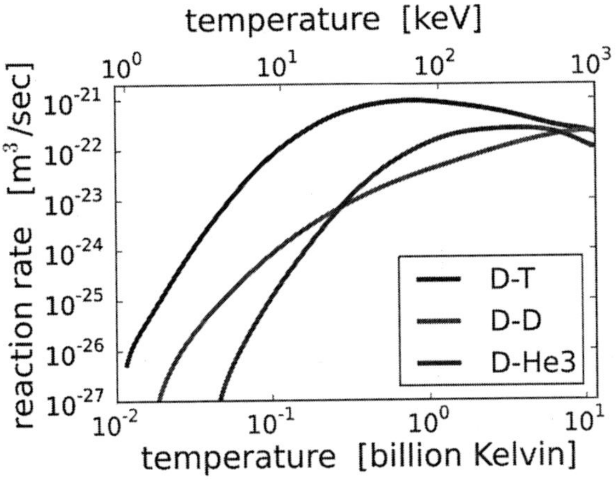

Figure 12: The fusion reaction rate increases rapidly with temperature until it maximises and then gradually drops off. The deuterium-tritium fusion rate peaks at a lower temperature (about 70 keV, or 800 million kelvin) and at a higher value than other reactions commonly considered for fusion energy.

The basic concept behind any fusion reaction is to bring two or more nuclei close enough together so that the residual strong force (nuclear force) in their nuclei will pull them together into one larger nucleus. If two light nuclei fuse, they will generally form a single nucleus with a slightly smaller mass than the sum of their original masses (though this is not always the case). The difference in mass is released as energy according to Albert Einstein's mass-energy equivalence formula $E = mc^2$. If the input nuclei are sufficiently massive, the resulting fusion product will be heavier than the sum of the reactants' original masses, in which case the reaction requires an external source of energy. The dividing line between "light" and "heavy" is iron-56. Above this atomic mass, energy will generally be released by nuclear fission reactions; below it, by fusion.

Fusion between the nuclei is opposed by their shared electrical charge, specifically the net positive charge of the protons in the nucleus. To overcome this electrostatic force, or "Coulomb barrier", some external source of energy must be supplied. The easiest way to do this is to heat the atoms, which has the side effect of stripping the electrons from the atoms and leaving them as bare nuclei. In most experiments the nuclei and electrons are left in a fluid known as a plasma. The temperatures required to provide the nuclei with enough energy to overcome their repulsion is a function of the total charge, so hydrogen, which has the smallest nuclear charge therefore reacts at the lowest temperature. Helium has an extremely low mass per nucleon and therefore is energetically favoured as a fusion product. As a consequence, most fusion reactions combine isotopes of hydrogen ("protium", deuterium, or tritium) to form isotopes of helium (^{3}He or ^{4}He).

The reaction cross section, denoted σ, is a measure of the probability of a fusion reaction as a function of the relative velocity of the two reactant nuclei. If the reactants have a distribution of velocities, as is the case in a thermal distribution within a plasma, then it is useful to perform an average over the distributions of the product of cross section and velocity. The reaction rate (fusions per volume per time) is $\langle\sigma v\rangle$ times the product of the reactant number densities:

$$f = (\tfrac{1}{2}n)^2 \langle\sigma v\rangle \text{ (for one reactant)}$$

$$f = n_1 n_2 \langle\sigma v\rangle \text{ (for two reactants)}$$

$\langle\sigma v\rangle$ increases from virtually zero at room temperatures up to meaningful magnitudes at temperatures of 10–100 keV (2.2–22 fJ). The significance of $\langle\sigma v\rangle$ as a function of temperature in a device with a particular energy confinement time is found by considering the Lawson criterion.

Perhaps the three most widely considered fuel cycles are based on the D-T, D-D, and p-^{11}B reactions. Other fuel cycles (D-^{3}He and ^{3}He-^{3}He) would require a supply of ^{3}He, either from other nuclear reactions or from extraterrestrial sources, such as the surface of the moon or the atmospheres of the gas giant planets. The details of the calculations comparing these reactions can be found here.

D-T Fuel Cycle

The easiest (according to the Lawson criterion) and most immediately promising nuclear reaction to be used for fusion power is:

$$^2_1\mathrm{D} + ^3_1\mathrm{T} \rightarrow ^4_2\mathrm{He} + ^1_0\mathrm{n}$$

Hydrogen-2 (Deuterium) is a naturally occurring isotope of hydrogen and as such is universally available. The large mass ratio of the hydrogen isotopes makes the separation rather easy compared to the difficult uranium enrichment process. Hydrogen-3 (Tritium) is also an isotope of hydrogen, but it occurs naturally in only negligible amounts due to its radioactive half-life of 12.32 years. Consequently, the deuterium-tritium fuel cycle requires the breeding of tritium from lithium using one of the following reactions:

$$^{1}_{0}n + ^{6}_{3}Li \rightarrow ^{3}_{1}T + ^{4}_{2}He$$

$$^{1}_{0}n + ^{7}_{3}Li \rightarrow ^{3}_{1}T + ^{4}_{2}He + ^{1}_{0}n$$

The reactant neutron is supplied by the D-T fusion reaction shown above, the one that also produces the useful energy. The reaction with ^{6}Li is exothermic, providing a small energy gain for the reactor. The reaction with ^{7}Li is endothermic but does not consume the neutron. At least some ^{7}Li reactions are required to replace the neutrons lost by reactions with other elements. Most reactor designs use the naturally occurring mix of lithium isotopes. However, the supply of lithium is relatively limited with other applications such as Li-ion batteries increasing its demand.

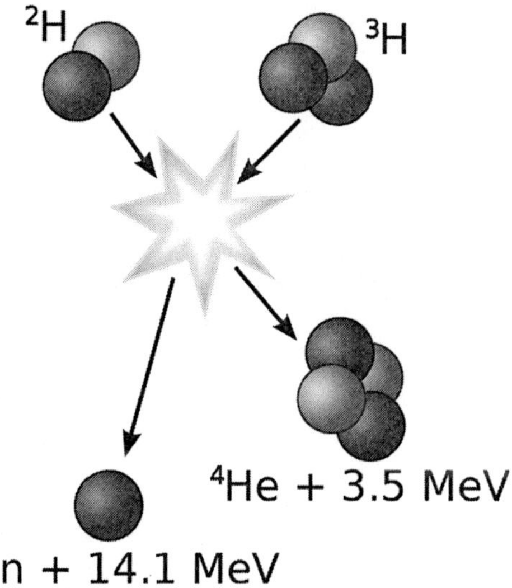

Figure 13: Diagram of the D-T reaction

Several drawbacks are commonly attributed to D-T fusion power:

1. It produces substantial amounts of neutrons that result in induced radioactivity within the reactor structure.
2. Only about 20% of the fusion energy yield appears in the form of charged particles (the rest neutrons), which limits the extent to which direct energy conversion techniques might be applied.

3. The use of D-T fusion power depends on lithium resources, which are less abundant than deuterium resources. However, lithium is relatively abundant on earth.
4. It requires the handling of the radioisotope tritium. Similar to hydrogen, tritium is difficult to contain and may leak from reactors in some quantity. Some estimates suggest that this would represent a fairly large environmental release of radioactivity.

The neutron flux expected in a commercial D-T fusion reactor is about 100 times that of current fission power reactors, posing problems for material design. Design of suitable materials is under way but their actual use in a reactor is not proposed until the generation after ITER. After a single series of D-T tests at JET, the largest fusion reactor yet to use this fuel, the vacuum vessel was sufficiently radioactive that remote handling needed to be used for the year following the tests.

In a production setting, the neutrons would be used to react with lithium in order to create more tritium. This also deposits the energy of the neutrons in the lithium, which would then be cooled to remove this energy and drive electrical production. This reaction protects the outer portions of the reactor from the neutron flux. Newer designs, the advanced tokamak in particular, also use lithium inside the reactor core as a key element of the design. The plasma interacts directly with the lithium, preventing a problem known as "recycling". The advantage of this layout was demonstrated in the Lithium Tokamak Experiment.

D-D Fuel Cycle

Though more difficult to facilitate than the deuterium-tritium reaction, fusion can also be achieved through the reaction of deuterium with itself. This reaction has two branches that occur with nearly equal probability:

$^2H + ^2H$	$\rightarrow$ ^3T	$+ ^1H$
	$\rightarrow$ ^3He	$+ n$

The optimum energy for this reaction is 15 keV, only slightly higher than the optimum for the D-T reaction. The first branch does not produce neutrons, but it does produce tritium, so that a D-D reactor will not be completely tritium-free, even though it does not require an input of tritium or lithium. Most of the tritium produced will be burned before leaving the reactor, which reduces the tritium handling required, but also means that more neutrons are produced and that some of these are very energetic.

The neutron from the second branch has an energy of only 2.45 MeV (0.393 pJ), whereas the neutron from the D-T reaction has an energy of 14.1 MeV (2.26 pJ), resulting in a wider range of isotope production and material damage. Assuming complete tritium burn-up, the reduction in the fraction of fusion energy carried by neutrons is only about 18%, so that the primary advantage of the D-D fuel cycle is that tritium breeding is not required. Other advantages are independence from limitations of lithium resources and a somewhat softer neutron spectrum. The price to pay compared to D-T is that the energy confinement (at a given pressure) must be 30 times better and the power produced (at a given pressure and volume) is 68 times less.

D-^{3}He Fuel Cycle

A second-generation approach to controlled fusion power involves combining helium-3 (^{3}He) and deuterium (^{2}H). This reaction produces a helium-4 nucleus (^{4}He) and a high-energy proton. As with the p-^{11}B aneutronic fusion fuel cycle, most of the reaction energy is released as charged particles, reducing activation of the reactor housing and potentially allowing more efficient energy harvesting (via any of several speculative technologies). In practice, D-D side reactions produce a significant number of neutrons, resulting in p-^{11}B being the preferred cycle for aneutronic fusion.

p-^{11}B Fuel Cycle

If aneutronic fusion is the goal, then the most promising candidate may be the Hydrogen-1 (proton)/boron reaction:

$$^1H + {}^{11}B \rightarrow 3\,{}^4He$$

Under reasonable assumptions, side reactions will result in about 0.1% of the fusion power being carried by neutrons. At 123 keV, the optimum temperature for this reaction is nearly ten times higher than that for the pure hydrogen reactions, the energy confinement must be 500 times better than that required for the D-T reaction, and the power density will be 2500 times lower than for D-T. Since the confinement properties of conventional approaches to fusion such as the tokamak and laser pellet fusion are marginal, most proposals for aneutronic fusion are based on radically different confinement concepts, such as the Polywell and the Dense plasma focus.

History of Research

The idea of using human-initiated fusion reactions was first made practical for military purposes in nuclear weapons. In a hydrogen bomb, the energy released by a fission weapon is used to compress and heat fusion fuel, beginning a fusion reaction that releases a large amount of neutrons that increases the rate of fission. The first fission-fusion-fission-based weapons released some 500 times more energy than early fission weapons.

Attempts at controlling fusion had already started by this point. Registration of the first patent related to a fusion reactor by the United Kingdom Atomic Energy Authority, the inventors being Sir George Paget Thomson and Moses Blackman, dates back to 1946. This was the first detailed examination of the pinch concept, and small efforts to experiment with the pinch concept started at several sites in the UK.

Around the same time, an expatriate German Ronald Richter proposed the Huemul Project in Argentina, announcing positive results in 1951. Although these results turned out to be false, it sparked off intense interest around the world. The UK pinch programs were greatly expanded, culminating in the ZETA and Sceptre devices. In the US, pinch experiments like those in the UK started at the Los Alamos National Laboratory. Similar devices were built in the USSR after data on the UK program was passed to them by Klaus Fuchs. At Princeton University a new approach developed as the stellarator, and the research establishment formed there continues to this day as the Princeton Plasma Physics Laboratory. Not to be outdone, Lawrence Liver more National Laboratory entered the field with their own variation, the magnetic mirror. These three groups have remained the primary developers of fusion research in the US to this day.

In the time since these early experiments, two new approaches developed that have since come to dominate fusion research. The first was the tokamak approach developed in the Soviet Union, which combined features of the stellarator and pinch to produce a device that dramatically outperformed either. The majority of magnetic fusion research to this day has followed the tokamak approach. In the late 1960s the concept of "mechanical" fusion through the use of lasers was developed in the US, and Lawrence Livermore switched their attention from mirrors to lasers over time.

Civilian applications are still being developed. Although it took less than ten years for fission to go from military applications to civilian fission energy production, it has been very different in the fusion energy field; more than fifty years have already passed since the first fusion reaction took place and sixty years since the first attempts to produce controlled fusion power, without any commercial fusion energy production plant coming into operation.

Magnetic Containment

Pinch Devices: A major area of study in early fusion power research is the "pinch" concept. Pinch is based on the fact the plasmas are electrically conducting. By running a current through the plasma, a magnetic field will be generated around the plasma. This field will, according to Lenz's law, create an inward directed force that causes the plasma to collapse inward, raising its density. Denser plasmas generate denser magnetic fields, increasing the inward force, leading to a chain reaction. If the conditions are correct, this can lead to the densities and temperatures needed for fusion. The trick is getting the current into the plasma; this is solved by inducing the current from an external magnet, which also produces the external field the internal field acts against.

Pinch was first developed in the UK in the immediate post-war era. Starting in 1947 small experiments were carried out and plans were laid to build a much larger machine. When the Huemul results hit the news, James L. Tuck, a UK physicist working at Los Alamos, introduced the pinch concept in the US and produced a series of machines known as the Perhapsatron. In the Soviet Union, a series of similar machines were being built, unknown in the west. All of these devices quickly demonstrated a series of instabilities in the fusion when the pinch was applied, which broke up the plasma column long before it reached the densities and temperatures needed for fusion. In 1953 Tuck and others suggested a number of solutions to these problems.

The largest "classic" pinch device was the ZETA, including all of these upgrades, starting operations in the UK in 1957. In early 1958 John Cockcroft announced that fusion had been achieved in the ZETA, an announcement that made headlines around the world. When physicists in the US expressed concerns about the claims they were initially dismissed. However, US experiments demonstrated the same neutrons, although measurements suggested these could not be from fusion reactions. The neutrons seen in the UK were later demonstrated to be from different versions of the same instability processes that plagued earlier machines. Cockcroft was forced to retract the fusion claims, which tainted the entire field for years. ZETA ended its experiments in 1968, and most other pinch experiments ended shortly after.

In 1974 a study of the ZETA results demonstrated an interesting side-effect; after the experimental runs ended, the plasma would enter a short period of stability. This led to the reversed field pinchconcept which has seen some level of development ever since. Recent work on the basic concept started as a result of the appearance of the "wires array" concept in the 1980s, which allowed a more efficient use of this technique. The Sandia National Laboratory runs a continuing wire-array research program with the Zpinch machine. In addition, the University of Washington's ZaP Lab has shown quiescent periods of stability hundreds of times longer than expected for plasma in a Z-pinch configuration, giving promise to the confinement technique.

In 1995, the staged Z-pinch concept was introduced by a team of scientist from University of California Irvine (UCI). This scheme can control one of the most dangerous instability that normally disintegrate conventional Z-pinch before the final implosion. The concept is based on a complex load of radiative liner plasma embedded with a target plasma. During implosion the outer surface of the liner plasma becomes unstable but the target plasma remains remarkably stable, up until the final implosion, generating a very high energy density stable target plasma. The heating mechanisms are shock heating, adiabatic compression and trapping of charge particles produced in fusion reaction due to a very strong magnetic field, which develops between the liner and the target. Details of this concept are shown in various publications available on the web page of MIFTI.

Early Magnetic Approaches

The U.S. fusion program began in 1951 when Lyman Spitzer began work on a stellarator under the code name Project Matterhorn. His work led to the creation of the Princeton Plasma Physics Laboratory, where magnetically confined plasmas are still studied. Spitzer planned an aggressive development project of four machines, A, B, C, and D. A and B were small research devices, C would be the prototype of a power-producing machine, and D would be the prototype of a commercial device. A worked without issue, but even by the time B was being used it was clear the stellarator was also suffering from instabilities and plasma leakage. Progress on C slowed as attempts were made to correct for these problems.

At Lawrence Livermore, the magnetic mirror was the preferred approach. The mirror consisted of two large magnets arranged so they had strong fields within them, and a weaker, but connected, field between them. Plasma introduced in the area between the two magnets would "bounce back" from the stronger fields in the middle. Although the design would leak plasma through the mirrors, the rate of leakage would be low enough that a useful fusion rate could be maintained. The simplicity of the design was supposed to make up for its lower performance. In practice the mirror also suffered from mysterious leakage problems, and never reached the expected performance.

Gun Club, MHD, Instability; Progress Slows

By the mid-1950s it was clear that the simple theoretical tools being used to calculate the performance of all fusion machines were simply not predicting their actual behaviour. Machines invariably leaked their plasma from their confinement area at rates far higher than predicted.

In 1954, Edward Teller held a gathering of fusion researchers at the Princeton Gun Club, near the Project Matterhorn (now known as Project Sherwood) grounds. Teller started by pointing out the problems that everyone was having, and suggested that any system where the plasma was confined within concave fields was doomed to fail. Attendees remember him saying something to the effect that the fields were like rubber bands, and they would attempt to snap back to a straight configuration whenever the power was increased, ejecting the plasma. He went on to say that it appeared the only way to confine the plasma in a stable configuration would be to use convex fields, a "cusp" configuration.

When the meeting concluded, most of the researchers quickly turned out papers saying why Teller's concerns did not apply to their particular device. The pinch machines did not use magnetic fields in this way at all, while the mirror and stellarator seemed to have various ways out. However, this was soon followed by a paper by Martin David Kruskal and Martin Schwarzschild discussing pinch machines, which demonstrated instabilities in those devices were inherent to the design. A series of similar studies followed, abandoning the simplistic theories previously used and introducing a full consideration of magnetohydrodynamics with a partially-resistive plasma. These concepts developed quickly, and by the early 1960s it was clear that small devices simply would not work. A series of much larger and more complex devices followed as researchers attempted to add field upon field in order to provide the required field strength without reaching the unstable regimes. As cost and complexity climbed, the initial optimism of the fusion field faded.

The Tokamak is Announced

A new approach was outlined in the theoretical works fulfilled in 1950–1951 by I.E. Tamm and A.D. Sakharov in the Soviet Union, which first discussed a tokamak-like approach. Experimental research on these designs began in 1956 at the Kurchatov Institute in Moscow by a group of Soviet scientists led by Lev Artsimovich. The tokamak essentially combined a low-power pinch device with a low-power simple stellarator. The key was to combine the fields in such a way that the particles wound around the reactor a particular number of times, today known as the "safety factor". The combination of these fields dramatically improved confinement times and densities, resulting in huge improvements over existing devices.

The group constructed the first tokamaks, the most successful being the T-3 and its larger version T-4. T-4 was tested in 1968 in Novosibirsk, producing the first quasistationary thermonuclear fusion reaction ever. The tokamak was dramatically more efficient than the other approaches of that era, on the order of 10 to 100 times. When they were first announced the international community was highly skeptical. However, a British team was invited to see T-3, and having measured it in depth they released their results that confirmed the Soviet claims. A burst of activity followed as many planned devices were abandoned and new tokamaks were introduced in their place - the C model stellarator, then under construction after many redesigns, was quickly converted to the Symmetrical Tokamakand the stellarator was abandoned.

Through the 1970s and 80s great strides in understanding the tokamak system were made. A number of improvements to the design are now part of the "advanced tokamak" concept, which includes non-circular plasmas, internal diverters and limiters, often superconducting magnets, and operate in the so-called "H-mode" island of increased stability. Two other designs have also become fairly well studied; the compact tokamak is wired with the magnets on the inside of the vacuum chamber, while the spherical tokamak reduces its cross section as much as possible.

The tokamak dominates modern research, where very large devices like ITER are expected to pass several milestones toward commercial power production, including a burning plasma with long burn times, high power output, and online fuelling. There are no guarantees that the project will be successful; previous generations of tokamak machines have uncovered new problems many times. But the entire field of high temperature plasmas is much better understood now than formerly, and there is considerable optimism that ITER will meet its goals. If successful, ITER would be followed by a "commercial demonstrator" system, similar in purpose to the very earliest power-producing fission reactors built in the era before wide-scale commercial deployment of larger machines started in the 1960s and 1970s.

Even with these goals met, there are a number of major engineering problems remaining, notably finding suitable "low activity" materials for reactor construction, demonstrating secondary systems including practical tritium extraction, and building reactor designs that allow their reactor core to be removed when its materials becomes embrittled due to the neutron flux. Practical commercial generators based on the tokamak concept are far in the future. The public at large has been disappointed, as the initial outlook for practical fusion power plants was much rosier; a pamphlet from the 1970s printed by General Atomic stated that "Several commercial fusion reactors are expected to be online by the year 2000."

Inertial (Laser) Containment

The technique of implosion of a microcapsule irradiated by laser beams, the basis of laser inertial confinement, was first suggested in 1962 by scientists at Lawrence Livermore National Laboratory, shortly after the invention of the laser itself in 1960. Lasers of the era were very low powered, but low-level research using them nevertheless started as early as 1965. A great advance in the field was John Nuckolls' 1972 paper that ignition would require lasers of about 1 kJ, and efficient burn around 1 MJ. kJ lasers were just beyond the state of the art at the time, and his paper sparked off a tremendous development effort to produce devices of the needed power.

Early machines used a variety of approaches to attack one of two problems - some focused on fast delivery of energy, while others were more interested in beam smoothness. Both were attempts to ensure the energy delivery would be smooth enough to cause an even implosion. However, these experiments demonstrated a serious problem; laser wavelengths in the infrared area lost a tremendous amount of energy before compressing the fuel. Important breakthroughs in this laser technology were made at the Laboratory for Laser Energetics at the University of Rochester, where scientists used frequency-tripling crystals to transform the infrared laser beams into ultraviolet beams. By the late 1970s great strides had been made in laser power, but with each increase new problems were found in the implosion technique that suggested even more power would be required. By the 1980s these increases were so large that using the concept for generating net energy seemed remote. Most research in this field turned to weapons research, always a second line of research, as the implosion concept is somewhat similar to hydrogen bomb operation. Work on very large versions continued as a result, with the very large National Ignition Facility in the US and Laser Mégajoule in France supporting these research programs.

More recent work had demonstrated that significant savings in the required laser energy are possible using a technique known as "fast ignition". The savings are so dramatic that the concept appears to be a useful technique for energy production again, so much so that it is a serious contender for pre-commercial development. There are proposals to build an experimental facility dedicated to the fast ignition approach, known as HiPER. At the same time, advances in solid state lasers appear to improve the "driver" systems' efficiency by about ten times (to 10- 20%), savings that make even the large "traditional" machines almost practical, and might make the fast ignition concept outpace the magnetic approaches in further development.

The laser-based concept has other advantages. The reactor core is mostly exposed, as opposed to being wrapped in a huge magnet as in the tokamak. This makes the problem of removing energy from the system somewhat simpler, and should mean that a laser-based device would be much easier to perform maintenance on, such as core replacement. Additionally, the lack of strong magnetic fields allows for a wider variety of low-activation materials, including carbon fibre, which would reduce both the frequency of such neutron activations and the rate of irradiation to the core. In other ways the program has many of the same problems as the tokamak; practical methods of energy removal and tritium recycling need to be demonstrated.

Other Approaches

Over the years there have been a wide variety of fusion concepts. In general they fall into three groups - those that attempt to reach high temperature/density for brief times (pinch, inertial confinement), those that operate at a steady state (magnetic confinement) or those that try neither and instead attempt to produce low quantities of fusion but do so at an extremely low cost. The latter group has largely disappeared, as the difficulties of achieving fusion have demonstrated that any low-energy device is unlikely to produce net gain. This leaves the two major approaches, magnetic and laser inertial, as the leading systems for development funding. However, alternate approaches continue to be developed, and alternate non-power fusion devices have been successfully developed as well.

Technically Viable

Philo T. Farnsworth, the inventor of the first all-electronic television system in 1927, patented his first Fusor design in 1968, a device that uses inertial electrostatic confinement. This system consists largely of two concentric spherical electrical grids inside a vacuum chamber into which a small amount of fusion fuel is introduced. Voltage across the grids causes the fuel to ionise around them, and positively charged ions are accelerated towards the centre of the chamber. Those ions may collide and fuse with ions coming from the other direction, may scatter without fusing, or may pass directly through. In the latter two cases, the ions will tend to be stopped by the electric field and re-accelerated toward the centre. Fusors can also use ion guns rather than electric grids. Towards the end of the 1960s, Robert Hirsch designed a variant of the Farns worth Fusor known as the Hirsch-Meeks fusor. This variant is a considerable improvement over the Farns worth design, and is able to generate neutron flux in the order of one billion neutrons per second. Although the efficiency was very low at first, there were hopes the device could be scaled up, but continued development demonstrated that this approach would be impractical for large machines. Nevertheless, fusion could be achieved using a "lab bench top" type set up for the first time, at minimal cost. This type of fusor found its first application as a portable neutron generator in the late 1990s. An automated sealed reaction chamber version of this device, commercially named Fusionstar was developed by EADS but abandoned in 2001. Its successor is the NSD-Fusion neutron generator.

Robert W. Bussard's Polywell concept is roughly similar to that of the Fusor, but replaces the problematic grid with a magnetically contained electron cloud, which holds the ions in position and provides an accelerating potential. The polywell consists of electromagnet coils arranged in a polyhedral configuration and positively charged to between several tens and low hundreds of kilovolts. This charged magnetic polyhedron is called a MaGrid (Magnetic Grid). Electrons are introduced outside the "quasi-spherical" MaGrid and are accelerated into the MaGrid due to the electric field, similar to a magnetic bottle. Within the MaGrid, magnetic fields confine most of the electrons and those that escape are retained by the electric field. This configuration traps the electrons in the middle of the device, focusing them near the centre to produce a virtual cathode (negative electric potential). The virtual cathode accelerates and confines the ions to be fused which, except for minimal losses, never reach the physical structure of the MaGrid. Bussard had reported a fusion rate of 10^9 per second running D-D fusion reactions at only 12.5 kV based on detecting a total of nine neutrons in five tests. Bussard claimed a scaled-up version of 2.5–3 m in diameter, would operate at over 100 MW net power since fusion power scales as the fourth power of the B field and the cube of the size.

A recent area of study is the magneto-inertial fusion (MIF) concept, which combines some form of external inertial compression (like lasers) with further compression through an external magnet (like pinch devices). The magnetic field traps heat within the inertial core, causing a variety of effects that improves fusion rates. These improvements are relatively minor, however the magnetic drivers themselves are inexpensive compared to lasers or other systems. There is hope for a sweet spot that allows the combination of features from these devices to create low-density but also low-cost fusion devices. A similar concept is the magnetised target fusion device, which uses a magnetic field in an external metal shell to achieve the same basic goals.

According to Eric Lerner, Focus fusion takes place in a dense plasma focus, which typically consists of two coaxial cylindrical electrodes made from copper or beryllium and housed in a vacuum chamber containing a low-pressure gas, which is used as the reactor fuel. An electrical pulse is applied across the electrodes, producing heating and a magnetic field. The current forms the hot gas into many minuscule vortices perpendicular to the surfaces of the electrodes, which then migrate to the end of the inner electrode to pinch-and-twist off as tiny balls of plasma called plasmoids. The electron beam collides with the plasmoid, heating it to fusion temperatures. This will, in principle, yield more energy in the beams than was input to form them.

Non-power Generating Approaches

A more subtle technique is to use more unusual particles to catalyse fusion. The best known of these is Muon-catalysed fusion which uses muons, which behave somewhat like electrons and replace the electrons around the atoms. These muons allow atoms to get much closer and thus reduce the kinetic energy required to initiate fusion. Muons require more energy to produce than can be obtained from muon-catalysed fusion, making this approach impractical for the generation of power.

In April 2005, a team from UCLA announced it had devised a way of producing fusion using a machine that "fits on a lab bench", using lithium tantalate to generate enough voltage to smash deuterium atoms together. However, the process does not generate net power. Such a device would be useful in the same sort of roles as the fusor.

Unconventional Approaches

Some scientists reported excess heat, neutrons, tritium, helium and other nuclear effects in so-called cold fusion systems, which for a time gained interest as showing promise. Hopes fell when replication failures were weighed in view of several reasons cold fusion is not likely to occur, the discovery of possible sources of experimental error, and finally the discovery that Fleischmann and Pons had not actually detected nuclear reaction by-products. By late 1989, most scientists considered cold fusion claims dead, and cold fusion subsequently gained a reputation as pathological science. However, a small community of researchers continues to investigate cold fusion claiming to replicate Fleishmann and Pons' results including nuclear reaction by-products. Claims related to cold fusion are largely disbelieved in the mainstream scientific community. In 1989, the majority of a review panel organised by the US Department of Energy (DOE) found that the evidence for the discovery of a new nuclear process was not persuasive. A second DOE review, convened in 2004 to look at new research, reached conclusions similar to the first.

Research into sonolumine scence induced fusion, sometimes known as "*bubble fusion*", also continues, although it is met with as much skepticism as cold fusion is by most of the scientific community.

Key Design Areas

Confinement : Confinement refers to all the conditions necessary to keep a plasma dense and hot long enough to undergo fusion:

- Equilibrium: There must be no net forces on any part of the plasma, otherwise it will rapidly disassemble. The exception, of course, is inertial confinement, where the relevant physics must occur faster than the disassembly time.

- Stability: The plasma must be so constructed that small deviations are restored to the initial state, otherwise some unavoidable disturbance will occur and grow exponentially until the plasma is destroyed.
- Transport: The loss of particles and heat in all channels must be sufficiently slow. The word "confinement" is often used in the restricted sense of "energy confinement".

The first human-made, large-scale fusion reaction was the test of the hydrogen bomb, Ivy Mike, in 1952. As part of the PACER project, it was once proposed to use hydrogen bombs as a source of power by detonating them in underground caverns and then generating electricity from the heat produced, but such a power plant is unlikely ever to be constructed, for a variety of reasons. *Controlled* thermonuclear fusion (CTF) refers to the alternative of continuous power production, or at least the use of explosions that are so small that they do not destroy a significant portion of the machine that produces them.

To produce self-sustaining fusion, the energy released by the reaction (or at least a fraction of it) must be used to heat new reactant nuclei and keep them hot long enough that they also undergo fusion reactions. Retaining the heat is called energy confinement and may be accomplished in a number of ways.

The hydrogen bomb really has no confinement at all. The fuel is simply allowed to fly apart, but it takes a certain length of time to do this, and during this time fusion can occur. This approach is called inertial confinement. If more than milligram quantities of fuel are used (and efficiently fused), the explosion would destroy the machine, so theoretically, controlled thermonuclear fusion using inertial confinement would be done using tiny pellets of fuel which explode several times a second. To induce the explosion, the pellet must be compressed to about 30 times solid density with energetic beams. If the beams are focused directly on the pellet, it is called direct drive, which can in principle be very efficient, but in practice it is difficult to obtain the needed uniformity. An alternative approach is indirect drive, in which the beams heat a shell, and the shell radiates x-rays, which then implode the pellet. The beams are commonly laser beams, but heavy and light ion beams and electron beams have all been investigated.

Inertial confinement produces plasmas with impressively high densities and temperatures, and appears to be best suited to weapons research, X-ray generation, very small reactors, and perhaps in the distant future, spaceflight. They require fuel pellets with close to a perfect shape in order to generate a symmetrical inward shock wave to produce the high-density plasma, and in practice these have proven difficult to produce. A recent development in the field of laser induced ICF is the use of ultrashort pulse multi-petawatt lasers to heat the plasma of an imploding pellet at exactly the moment of greatest density after it is imploded conventionally using terawatt scale lasers. This research will be carried out on the (currently being built) OMEGA EP petawatt and OMEGA lasers at the University of Rochester and at the GEKKO XII laser at the institute for laser engineering in Osaka Japan, which if fruitful, may have the effect of greatly reducing the cost of a laser fusion based power source.

At the temperatures required for fusion, the fuel is in the form of a plasma with very good electrical conductivity. This opens the possibility to confine the fuel and the energy with magnetic fields, an idea known as magnetic confinement. The Lorenz force works only perpendicular to the magnetic field, so that the first problem is how to prevent the plasma from leaking out the ends of the field lines. There are basically two solutions.

The first is to use the magnetic mirror effect. If particles following a field line encounter a region of higher field strength, then some of the particles will be stopped and reflected. Advantages of a magnetic mirror power plant would be simplified construction and maintenance due to a linear topology and the potential to apply direct conversion in a natural way, but the confinement achieved in the experiments was so poor that this approach has been essentially abandoned.

The second possibility to prevent end losses is to bend the field lines back on themselves, either in circles or more commonly in nested toroidal surfaces. The most highly developed system of this type is the *tokamak*, with the *stellarator* being next most advanced, followed by the Reversed field pinch. Compact toroids, especially the *Field-Reversed Configuration* and the spheromak, attempt to combine the advantages of toroidal magnetic surfaces with those of a simply connected (non-toroidal) machine, resulting in a mechanically simpler and smaller confinement area. Compact toroids still have some enthusiastic supporters but are not backed as readily by the majority of the fusion community.

Finally, there are also *electrostatic confinement fusion* systems, in which ions in the reaction chamber are confined and held at the centre of the device by electrostatic forces, as in the Farns worth-Hirsch Fusor, which is not believed to be able to be developed into a power plant. The Polywell, an advanced variant of the fusor, has shown a degree of research interest as of late; however, the technology is relatively immature, and major scientific and engineering questions remain which researchers under the auspices of the U.S. Office of Naval Research hope to further investigate.

Materials

Developing materials for fusion reactors has long been recognised as a problem nearly as difficult and important as that of plasma confinement, but it has received only a fraction of the attention. The neutron flux in a fusion reactor is expected to be about 100 times that in existing pressurised water reactors (PWR). Each atom in the blanket of a fusion reactor is expected to be hit by a neutron and displaced about a hundred times before the material is replaced. Furthermore the high-energy neutrons will produce hydrogen and helium in various nuclear reactions that tends to form bubbles at grain boundaries and result in swelling, blistering or embrittlement. One also wishes to choose materials whose primary components and impurities do not result in long-lived radioactive wastes. Finally, the mechanical forces and temperatures are large, and there may be frequent cycling of both.

The problem is exacerbated because realistic material tests must expose samples to neutron fluxes of a similar level for a similar length of time as those expected in a fusion power plant. Such a neutron source is nearly as complicated and expensive as a fusion reactor itself would be. Proper materials testing will not be possible in ITER, and a proposed materials testing facility, IFMIF, was still at the design stage in 2005.

The material of the plasma facing components (PFC) is a special problem. The PFC do not have to withstand large mechanical loads, so neutron damage is much less of an issue. They do have to withstand extremely large thermal loads, up to 10 MW/m^2, which is a difficult but solvable problem. Regardless of the material chosen, the heat flux can only be accommodated without melting if the distance from the front surface to the coolant is not more than a centimetre or two. The primary issue is the interaction with the plasma. One can choose either a low-Z material, typified by graphiteal though for some purposes beryllium might be chosen, or a high-Z material, usually tungsten with molybdenum as a second choice. Use of liquid metals (lithium, gallium, tin) has also been proposed, e.g., by injection of 1–5 mm thick streams flowing at 10 m/s on solid substrates.

If graphite is used, the gross erosion rates due to physical and chemical sputtering would be many meters per year, so one must rely on redeposition of the sputtered material. The location of the redeposition will not exactly coincide with the location of the sputtering, so one is still left with erosion rates that may be prohibitive. An even larger problem is the tritium co-deposited with the redeposited graphite. The tritium inventory in graphite layers and dust in a reactor could quickly build up to many kilograms, representing a waste of resources and a serious radiological hazard in case of an accident. The consensus of the fusion community seems to be that graphite, although a very attractive material for fusion experiments, cannot be the primary PFC material in a commercial reactor.

The sputtering rate of tungsten can be orders of magnitude smaller than that of carbon, and tritium is not so easily incorporated into redeposited tungsten, making this a more attractive choice. On the other hand, tungsten impurities in a plasma are much more damaging than carbon impurities, and self-sputtering of tungsten can be high, so it will be necessary to ensure that the plasma in contact with the tungsten is not too hot (a few tens of eV rather than hundreds of eV). Tungsten also has disadvantages in terms of eddy currents and melting in off-normal events, as well as some radiological issues.

Subsystems

In fusion research, achieving a fusion energy gain factor $Q = 1$ is called breakeven and is considered a significant although somewhat artificial milestone. Ignition refers to an infinite Q, that is, a self-sustaining plasma where the losses are made up for by fusion power without any external input. In a practical fusion reactor, some external power will always be required for things like current drive, refuelling, profile control, and burn control. A value on the order of $Q = 20$ will be required if the plant is to deliver much more energy than it uses internally.

Despite many differences between possible designs of power plant, there are several systems that are common to most. A fusion power plant, like a fission power plant, is customarily divided into the nuclear island and the balance of plant. The balance of plant converts heat into electricity via steam turbines; it is a conventional design area and in principle similar to any other power station that relies on heat generation, whether fusion, fission or fossil fuel based.

The nuclear island has a plasma chamber with an associated vacuum system, surrounded by plasma-facing components (first wall and divertor) maintaining the vacuum boundary and absorbing the thermal radiation coming from the plasma, surrounded in turn by a blanket where the neutrons are absorbed to breed tritium and heat a working fluid that transfers the power to the balance of plant. If magnetic confinement is used, a magnet system, using primarily cryogenic superconducting magnets, is needed, and usually systems for heating and refuelling the plasma and for driving current. In inertial confinement, a driver (laser or accelerator) and a focusing system are needed, as well as a means for forming and positioning the pellets.

Although the standard solution for electricity production in fusion power plant designs is conventional steam turbines using the heat deposited by neutrons, there are also designs for direct conversion of the energy of the charged particles into electricity. These are of little value with a D-T fuel cycle, where 80% of the power is in the neutrons, but are indispensable with aneutronic fusion, where less than 1% is. Direct conversion has been most commonly proposed for open-ended magnetic configurations like magnetic mirrors or Field-Reversed Configurations, where charged particles are lost along the magnetic field lines, which are then expanded to convert a large fraction of the random energy of the fusion products into directed motion. The particles are then collected on electrodes at various large electrical potentials. Typically the claimed conversion efficiency is in the range of 80%, but the converter may approach the reactor itself in size and expense.

Safety and the Environment

Accident Potential: There is no possibility of a *catastrophic* accident in a fusion reactor resulting in major release of radioactivity to the environment or injury to non-staff, unlike modern fission reactors. The primary reason is that nuclear fusion requires precisely controlled temperature, pressure, and magnetic field parameters to generate net energy. If the reactor were damaged, these parameters would be disrupted and the heat generation in the reactor would rapidly cease. Fusion reactors are extremely safe in this sense, and it makes them favourable over fission reactors, which, in contrast, continues to generate heat through beta-decay for several hours or even days after reactor shut-down, meaning that melting of fuel rods is possible even after the reactor has been stopped due to continued accumulation of heat.

There is also no risk of a runaway reaction in a fusion reactor, since the plasma is normally burnt at optimal conditions, and any significant change will render it unable to produce excess heat. In fusion reactors the reaction process is so delicate that this level of safety is inherent; no elaborate failsafe mechanism is required.

Although the plasma in a fusion power plant will have a volume of 1000 cubic meters or more, the density of the plasma is extremely low, and the total amount of fusion fuel in the vessel is very small, typically a few grams. If the fuel supply is closed, the reaction stops within seconds. In comparison, a fission reactor is typically loaded with enough fuel for one or several years, and no additional fuel is necessary to keep the reaction going.

In the magnetic approach, strong fields are developed in coils that are held in place mechanically by the reactor structure. Failure of this structure could release this tension and allow the magnet to "explode" outward.

The severity of this event would be similar to any other industrial accident or an MRI machine quench/explosion, and could be effectively stopped with a containment building similar to those used in existing (fission) nuclear generators. The laser-driven inertial approach is generally lower-stress. Although failure of the reaction chamber is possible, simply stopping fuel delivery would prevent any sort of catastrophic failure.

Most reactor designs rely on the use of liquid lithium as both a coolant and a method for converting stray neutrons from the reaction into tritium, which is fed back into the reactor as fuel. Lithium is highly flammable, and in the case of a fire it is possible that the lithium stored on-site could be burned up and escape. In this case the tritium contents of the lithium would be released into the atmosphere, posing a radiation risk.

However, calculations suggest that the total amount of tritium and other radioactive gases in a typical power plant would be so small, about 1 kg, that they would have diluted to legally acceptable limits by the time they blew as far as the plant's perimeter fence. The likelihood of *small industrial* accidents including the local release of radioactivity and injury to staff cannot be estimated yet. These would include accidental releases of lithium, tritium, or mis-handling of decommissioned radioactive components of the reactor itself.

Effluents During Normal Operation

The natural product of the fusion reaction is a small amount of helium, which is completely harmless to life. Of more concern is tritium, which, like other isotopes of hydrogen, is difficult to retain completely. During normal operation, some amount of tritium will be continually released. There would be no acute danger, but the cumulative effect on the world's population from a fusion economy could be a matter of concern.

Although tritium is volatile and biologically active, the health risk posed by a release is much lower than that of most radioactive contaminants, due to tritium's short half-life (12 years), very low decay energy (~14.95 keV), and the fact that it does not bioaccumulate (instead being cycled out of the body as water, with a biological half-life of 7 to 14 days). Current ITER designs are investigating total containment facilities for any tritium.

Waste Management

The large flux of high-energy neutrons in a reactor will make the structural materials radioactive. The radioactive inventory at shut-down may be comparable to that of a fission reactor, but there are important differences.

The half-life of the radioisotopes produced by fusion tend to be less than those from fission, so that the inventory decreases more rapidly. Unlike fission reactors, whose waste remains radioactive for thousands of years, most of the radioactive material in a fusion reactor would be the reactor core itself, which would be dangerous for about 50 years, and low-level waste another 100. Although this waste will be considerably more radioactive during those 50 years than fission waste, the very short half-life makes the process very attractive, as the waste management is fairly straightforward. By 300 years the material would have the same radioactivity as coal ash.

Additionally, the choice of materials used in a fusion reactor is less constrained than in a fission design, where many materials are required for their specific neutron cross-sections. This allows a fusion reactor to be designed using materials that are selected specifically to be "low activation", materials that do not easily become radioactive. Vanadium, for example, would become much less radioactive than stainless steel. Carbon fibre materials are also low-activation, as well as being strong and light, and are a promising area of study for laser-inertial reactors where a magnetic field is not required.

In general terms, fusion reactors would create far less radioactive material than a fission reactor, the material it would create is less damaging biologically, and the radioactivity "burns off" within a time period that is well within existing engineering capabilities.

Nuclear Proliferation

Although fusion power uses nuclear technology, the overlap with nuclear weapons technology is small. Tritium is a component of the trigger of hydrogen bombs, but not a major problem in production. The copious neutrons from a fusion reactor could be used to breed plutonium for an atomic bomb, but not without extensive redesign of the reactor, so that production would be difficult to conceal. The theoretical and computational tools needed for hydrogen bomb design are closely related to those needed for inertial confinement fusion, but have very little in common with the more scientifically developed magnetic confinement fusion.

As a Sustainable Energy Source

Large-scale reactors using neutronic fuels (e.g. ITER) and thermal power production (turbine based) are most comparable to fission power from an engineering and economics viewpoint. Both fission and fusion power plants involve a relatively compact heat source powering a conventional steam turbine-based power plant, while producing enough neutron radiation to make activation of the plant materials problematic.

The main distinction is that fusion power produces no high-level radioactive waste (though activated plant materials still need to be disposed of). There are some power plant ideas which may significantly lower the cost or size of such plants; however, research in these areas is nowhere near as advanced as in tokamaks.

Fusion power commonly proposes the use of deuterium, an isotope of hydrogen, as fuel and in many current designs also use lithium. Assuming a fusion energy output equal to the 1995 global power output of about 100 EJ/yr (= 1×10^{20} J/yr) and that this does not increase in the future, then the known current lithium reserves would last 3000 years, lithium from sea water would last 60 million years, and a more complicated fusion process using only deuterium from sea water would have fuel for 150 billion years. To put this in context, 150 billion years is over ten times the currently measured age of the universe, and is close to 30 times the remaining lifespan of the sun.

Economics

While fusion power is still in early stages of development, substantial sums have been and continue to be invested in research. In the EU almost € 10 billion was spent on fusion research up to the end of the 1990s, and the new ITER reactor alone is budgeted at € 10 billion. It is estimated that up to the point of possible implementation of electricity generation by nuclear fusion, R&D will need further promotion totalling around € 60-80 billion over a period of 50 years or so (of which € 20-30 billion within the EU). Nuclear fusion research receives € 750 million (excluding ITER funding), compared with € 810 million for all non-nuclear energy research combined, putting research into fusion power well ahead of that of any single rivalling technology.

Advantages

Fusion power would provide much more energy for a given weight of fuel than any technology currently in use, and the fuel itself (primarily deuterium) exists abundantly in the Earth's ocean: about 1 in 6500 hydrogen atoms in seawater is deuterium. Although this may seem a low proportion (about 0.015%), because nuclear fusion reactions are so much more energetic than chemical combustion and seawater is easier to access and more plentiful than fossil fuels, fusion could potentially supply the world's energy needs for millions of years.

Despite being technically non-renewable, fusion power has many of the benefits of renewable energy sources (such as being a long-term energy supply and emitting no greenhouse gases) as well as some of the benefits of the resource-limited energy sources as hydrocarbons and nuclear fission (without reprocessing). Like these currently dominant energy sources, fusion could provide very high power-generation density and uninterrupted power delivery (due to the fact that it is not dependent on the weather, unlike wind and solar power).

Another aspect of fusion energy is that the cost of production does not suffer from diseconomies of scale. The cost of water and wind energy, for example, goes up as the optimal locations are developed first, while further generators must be sited in less ideal conditions. With fusion energy, the production cost will not increase much, even if large numbers of plants are built.

Some problems which are expected to be an issue in this century such as fresh water shortages can alternatively be regarded as problems of energy supply. For example, in desalination plants, seawatercan be purified through distillation or reverse osmosis. However, these processes are energy intensive. Even if the first fusion plants are not competitive with alternative sources, fusion could still become competitive if large scale desalination requires more power than the alternatives are able to provide.

A scenario has been presented of the effect of the commercialisation of fusion power on the future of human civilization. ITER and later Demo are envisioned to bring online the first commercial nuclear fusion energy reactor by 2050. Using this as the starting point and the history of the uptake of nuclear fission reactors as a guide, the scenario depicts a rapid take up of nuclear fusion energy starting after the middle of this century.

Current Status

Despite optimism dating back to the 1950s about the wide-scale harnessing of fusion power, there are still significant barriers standing between current scientific understanding and technological capabilities and the practical realisation of fusion as an energy source. Research, while making steady progress, has also continually thrown up new difficulties. Therefore it remains unclear whether an economically viable fusion plant is possible. A 2006 editorial in *New Scientist* magazine opined that "if commercial fusion is viable, it may well be a century away.". This pessimistic view is in contrast to the optimism of a pamphlet printed by General Atomics in 1970s stated that "By the year 2000, several commercial fusion reactors are expected to be on-line."

Several fusion D-T burning tokamak test devices have been built (TFTR, JET), but these were not built to produce more thermal energy than electrical energy consumed. The ITER project is currently leading the effort to commercialise fusion power.

A paper published in January 2009 and part of the IAEA Fusion Conference Proceedings at Geneva last October claims that small 50 MW Tokamak style reactors are feasible.

On May 30, 2009, the US Lawrence Livermore National Laboratory (LLNL), announced the creation of a high-energy laser system, the National Ignition Facility, which can heat hydrogen atoms to temperatures only existing in nature in the cores of stars. The new laser is expected to have the ability to produce, for the first time, more energy from controlled, inertially confined nuclear fusion than was required to initiate the reaction.

On January 28, 2010, the LLNL announced tests using all 192 laser beams, although with lower laser energies, smaller hohlraum targets, and substitutes for the fusion fuel capsules. More than one megajoule of ultraviolet energy was fired into the hohlraum, beating the previous world record by a factor of more than 30. The results gave the scientists confidence that they will be able to achieve ignition in more realistic tests scheduled to begin in the summer of 2010.

NIF researchers are currently conducting a series of "tuning" shots to determine the optimal target design and laser parameters for high-energy ignition experiments with fusion fuel in the coming months. Two firing tests were performed on October 31 2010 and November 2 2010.

Electric Arc

An electric arc is an electrical breakdown of a gas which produces an ongoing plasma discharge, resulting from a current flowing through normally nonconductive media such as air. A synonym is arc discharge. An arc discharge is characterised by a lower voltage than a glow discharge, and relies on thermionic emission of electrons from the electrodes supporting the arc. An archaic term is voltaic arc as used in the phrase "voltaic arc lamp".

History

The phenomenon was first described in 1802, as a "special fluid with electrical properties", by Vasily V. Petrov, a Russian scientist experimenting with a copper-zinc battery consisting of 4200 discs. Sir Humphry Davy first demonstrated the arc early in the nineteenth century by transmitting an electric current through two touching carbon rods and then pulling them a short distance apart. In 1801, at a lecture before the Royal Society, he produced a "feeble" arc, not readily distinguished from a sustained spark, between charcoal points. The Society subscribed for a more powerful battery of 1000 plates and in 1808 he demonstrated the large-scale arc. He is credited with naming the arc.

Overview

The various shapes of electric arc are emergent properties of nonlinear patterns of current and electric field. The arc occurs in the gas-filled space between two conductive electrodes (often made of tungsten or carbon) and it results in a very high temperature, capable of melting or vaporising most materials. An electric arc is a continuous discharge, while a similar electric spark discharge is momentary. An electric arc may occur either in Direct current circuits or in alternating current circuits. In the latter case, the arc may re-strike on each half cycle of the current. An electric arc differs from a glow discharge in that the current density is quite high, and the voltage drop within the arc is low; at the cathode the current density may be as high as one mega ampere per square centimetre.

An electric arc has a non-linear relationship between current and voltage. Once the arc is established (either by progression from a glow discharge or by momentarily touching the electrodes then separating them), increased current results in a lower voltage between the arc terminals. This negative resistance effect requires that some positive form of impedance—an electrical ballast—to be placed in the circuit, to maintain a stable arc. This property is the reason uncontrolled electrical arcs in apparatus become so destructive, since once initiated an arc will draw more and more current from a fixed-voltage supply until the apparatus is destroyed.

Uses

Industrially, electric arcs are used for welding, plasma cutting, for electrical discharge machining, as an arc lampin movie projectors and follow spots in stage lighting. Electric arc furnaces are used to produce steel and other substances. Calcium carbide is made in this way as it requires a large amount of energy to promote anendothermic reaction (at temperatures of 2500 °C).

Low-pressure electric arcs are used for lighting, e.g., fluorescent tubes, mercury and sodium street lamps, and camera flash lamps.

Formation of an intense electric arc, similar to a small-scale arc flash, is the foundation of exploding-bridgewire detonators.

Electric arcs have been studied for electric propulsion of spacecraft.

Undesired Arcing

Undesired or unintended electric arcing can have detrimental effects on electric power transmission, distribution systems and electronic equipment. Devices which may cause arcing include switches, circuit breakers, relay contacts, fuses and poor cable terminations. When an inductive circuit is switched off the current cannot instantaneously jump to zero; a transient arc will be formed across the separating contacts. Switching devices susceptible to arcing are normally designed to contain and extinguish an arc, and snubber circuits can supply a path for transient currents, preventing arcing. If a circuit has enough current and voltage to sustain an arc formed outside of a switching device, the arc can cause damage to equipment such as melting of conductors, destruction of insulation, and fire. An arc flash describes an explosive electrical event that presents a hazard to people and equipment. Undesired arcing in electrical contactors can be suppressed by various devices, including:

- immersion in transformer oil, dielectric gas or vacuum
- arc chutes
- magnetic blowouts.

Arcing can also occur when a low resistance channel (foreign object, conductive dust, moisture...) forms between places with different potential. The conductive channel then can facilitate formation of an electric arc. The ionised air has high electrical conductivity approaching that of metals, and can conduct extremely high currents, causing a short circuit and tripping protective devices (fuses, circuit breakers). Similar situation may occur when a light bulb burns out and the fragments of the filament pull an electric arc between the leads inside the bulb, leading to overcurrent that trips the breakers.

Electric arc over the surface of plastics causes their degradation. A conductive carbon-rich track tends to form in the arc path, negatively influencing their insulation properties. The arc susceptibility is tested according to ASTM D495, by point electrodes and continuous and intermittent arcs; it is measured in seconds to form a track that is conductive under high-voltage low-current conditions. Some materials are less susceptible to degradation than others; e.g. polytetrafluoroethylene has arc resistance of about 200 seconds. From thermosetting plastics, alkyds and melamine resins are better than phenolic resins. Polyethylenes have arc resistance of about 150 seconds, polystyrenes and polyvinyl chlorides have relatively low resistance of about 70 seconds. Plastics can be formulated to emit gases with arc-extinguishing properties; these are known as *arc-extinguishing plastics.*

Arcing over some types of printed circuit boards, possibly due to cracks of the traces or the failure of a solder, renders the affected insulating layer conductive as the dielectric is combusted due to the high temperatures involved. This conductivity prolongs the arcing due to cascading failure of the surface.

The energy of electrical current contact arc yields ozone (O_3), along with other compounds such as nitrous oxides (NO, NO_x) and other chemicals and particulates. Unsuppressed arcing breaks down the chemical bonds of the atmospheric gases surrounding the contacts as well as some of the molten metal of the contact material itself. Free ions in and around the arc recombine to create new chemical compounds (for example, breaking atmospheric oxygen into single oxygen [O_2 '™ 2O], which then recombine creating ozone [O_3]. These chemicals and particulates are most pronounced when operating contactors, as they typically conduct higher contact power and run in the open. Ozone is most easily noticed during operation due to its distinct odour. Another environmental impact of undesired arcing is determined by the fact that the contact degradation drastically limits the overall life of a relay or conductor (the mechanical life of the device, which can be in excess of 20 million operations, is reduced to a range of about 10,000 to 100,000 operations), thus ending up much earlier in landfills and requiring faster replacement.

Arc Suppression

Arc suppression is a method of attempting to reduce to near elimination the electrical arc. There are several possible areas of use of arc suppression methods, among them metal film deposition and sputtering, arc flash protection, electrostatic processes where electrical arcs are not desired (such as powder painting, air purification, PVDF film poling) and contact current arc suppression. In industrial, military and consumer electronic design, the latter method generally applies to devices such as electromechanical power switches, relays and contactors. In this context, arc suppression refers to the concept of contact protection.

Plasma Globe

Plasma globes, or plasma lamps (also called plasma balls, domes, spheres, tubes or orbs, depending on shape), are novelty items that were most popular in the 1980s. The plasma lamp was invented by Nikola Tesla after his experimentation with high-frequency currents in an evacuated glass tube for the purpose of studying high voltage phenomena, but the modern versions were first designed by Bill Parker. Tesla called this invention an inert gas discharge tube.

Description

Most commonly, plasma globes are available in spheres or cylinders. Although many variations exist, a plasma lamp is usually a clear glass orb filled with a mixture of various gases (most commonly helium and neon, sometimes with other noble gases such as xenon and krypton) at low pressure (below 0.01 atmosphere) and driven by high-frequency alternating current at approximately 35 kHz,

2–5kV, generated by a high-voltage transformer. A much smaller orb in its centre serves as an electrode. Plasma filaments extend from the inner electrode to the outer glass insulator, giving the appearance of multiple constant beams of coloured light. The beams initially follow the electric field lines of the dipole but move upwards due to convection.

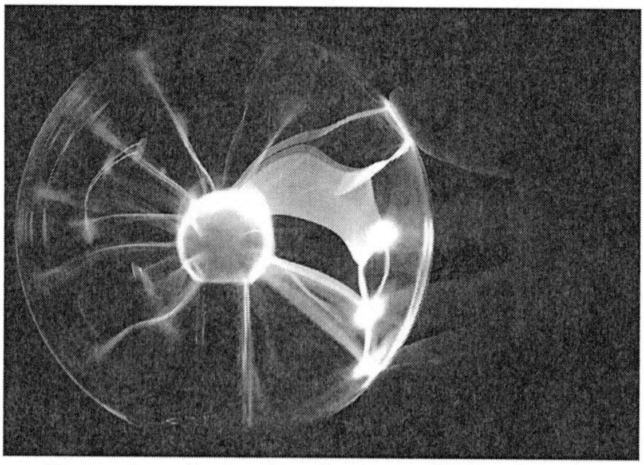

Figure 14: The effect of a conducting object (such as a hand) in close proximity with the plasma globe glass

Placing a hand near the glass alters the high-frequency electric field, causing a single beam to migrate from the inner ball to the point of contact. An electric current is produced within any conductive object near the orb, as the glass does not block the electromagnetic field created by the electric current flowing through the plasma (though the insulator does block the current itself). The glass acts as a dielectric in a capacitor formed between the ionised gas and the hand.

Potential Hazards

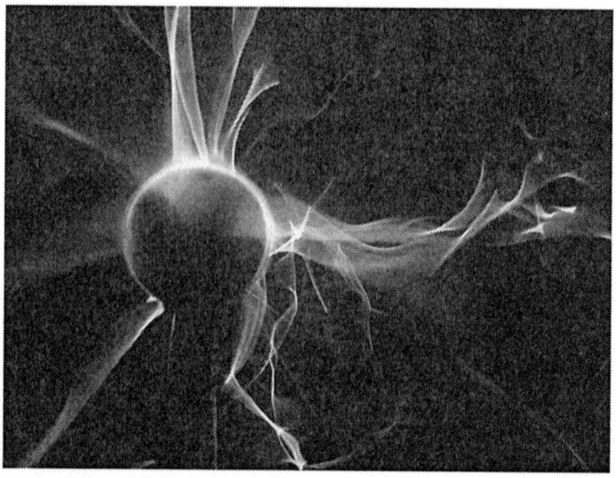

Figure 15: Interior of a plasma ball

Caution should be taken when placing electronic devices near or upon the plasma lamp: not only may the glass become hot, but the high voltage may place a substantial static charge on the device, even through a protective plastic casing. The radio frequency field produced by plasma lamps can interfere with the operation of touchpads used on laptop computers, digital audio players, cell phones, and other similar devices. Some types can radiate sufficient RFI to interfere with cordless telephones and Wi-Fi devices several feet away. If a medium-sized lamp is wrapped in grounded metal foil, capacitive couplingcan transfer tens of milliamperes to ground through the foil, enough to light a small lamp or give a small arc burn. This is possible because the glass acts as a capacitor dielectric: the inside of the lamp acts as one plate, and any conductive object on the outside acts as the other capacitor plate. Ozone, which is harmful to humans, may also accumulate outside of the surface of the glass orb after a few minutes of constant operation.

History

In U.S. Patent 0,514,170 ("Incandescent Electric Light", 1894 February 6), Nikola Tesla describes a plasma lamp. This patent is for one of the first high-intensity discharge lamps. Tesla used an incandescent-type lamp globe with a single internal conductive element and excited the element with high voltage currents from a Tesla coil, thus creating the brush discharge emanation. He gained patent protection on a particular form of the lamp in which a light-giving small body or button of refractory material is supported by a conductor entering a very highly exhausted globe or receiver. Tesla called this invention the single terminal lamp, or, later, the "Inert Gas Discharge Tube."

The popular product sold throughout the world today was invented by Bill Parker in 1970 as an undergraduate student at MIT.

The technology needed to formulate gas mixtures used in today's plasma spheres was not available to Tesla. Modern lamps typically use combinations of xenon, krypton and neon, although other gases can be used as well. These gas mixtures, along with different glass shapes and integrated-circuit-driven electronics, create the vivid colours, range of motions and complex patterns seen in today's plasma spheres.

Applications

Plasma lamps are mainly used as curiosities or toys for their unique lighting effects and the "tricks" that can be performed on them by users moving their hands around them. They might also form part of a school's laboratory equipment for demonstration purposes. They are not usually employed for general lighting. However, as of recent years, some novelty stores have begun selling a nightlight plasma lamp that can fit into a standard light socket.

Tesla Coil

A Tesla coil is a type of resonant transformer circuit invented by Nikola Tesla around 1891. It is used to produce high voltage, low current, high frequency alternating current electricity. Tesla coils produce higher current than the other source of high voltage discharges, electrostatic machines. Tesla experimented with a number of different configurations and they consist of two, or sometimes three, coupled resonant electric circuits. Tesla used these coils to conduct innovative experiments in electrical lighting, phosphorescence, x-ray generation, high frequency alternating current phenomena, electrotherapy, and the transmission of electrical energy without wires. Tesla coil circuits were used commercially in sparkgap radio transmitters for wireless telegraphy until the 1920s, and in pseudomedical equipment such as electrotherapy and violet ray devices. Today their main use is for entertainment and educational displays.

Theory

A Tesla coil transformer operates in a significantly different fashion from a conventional (i.e., iron core) transformer. In a conventional transformer, the windings are very tightly coupled and voltage gain is determined by the ratio of the numbers of turns in the windings. This works well at normal voltages but, at high voltages, the insulation between the two sets of windings is easily broken down and this prevents iron cored transformers from running at extremely high voltages without damage.

With Tesla coils, unlike a conventional transformer (which may couple 97%+ of the fields between windings) a Tesla coil's windings are "loosely" coupled, with a large air gap, and thus the primary and secondary typically share only 10–20% of their respective magnetic fields. Instead of a tight coupling, the coil transfers energy (via loose coupling) from one oscillating resonant circuit (the primary) to the other (the secondary) over a number of RF cycles.

As the primary energy transfers to the secondary, the secondary's output voltage increases until all of the available primary energy has been transferred to the secondary (less losses). Even with significant spark gap losses, a well designed Tesla coil can transfer over 85% of the energy initially stored in the primary capacitor to the secondary circuit. Thus the voltage gain of a Tesla coil can be significantly greater than a conventional transformer, since the air gap has a very high insulation.

With the loose coupling the voltage gain is instead proportional to the square root of the ratio of secondary and primary inductances.

History

The early Tesla coil transformer design employs a medium- to high-voltage power source, one or more high voltage capacitor(s), and a spark gap to excite a multiple-layer primary inductor with periodic bursts of high frequency current.

The multiple-layer Tesla coil transformer secondary is excited by resonant inductive coupling, the primary and secondary circuits both being *tuned* so they resonate at the same frequency (typically, between 25 kHz and 2 MHz). The later and higher-power coil design has a single-layer primary and secondary. These Tesla coils are often used by hobbyists and at venues such as science museums to produce long sparks. The "American Electrician" gives a description of an early Tesla coil wherein a glass battery jar, 15 x 20 cm (6 x 8 in) is wound with 60 to 80 turns of AWG No. 18 B & S magnet wire (0.823 mm^2). Into this is slipped a primary consisting of eight to ten turns of AWG No. 6 B & S wire (13.3 mm^2) and the whole combination immersed in a vessel containing linseed or mineral oil. (Norrie, pg. 34-35)

1902 Design

Tesla's 1902 design for his advanced magnifying transmitter used a top terminal consisting of a metal frame in the shape of a toroid, covered with hemispherical plates (constituting a very large conducting surface). The top terminal has relatively small capacitance, charged to as high a voltage as practicable. The outer surface of the elevated conductor is where the electrical charge chiefly accumulates. It has a large radius of curvature, or is composed of separate elements which, irrespective of their own radii of curvature, are arranged close to each other so that the outside ideal surface enveloping them has a large radius. This design allowed the terminal to support very high voltages without generating corona or sparks. Tesla, during his patent application process, described a variety of resonator terminals at the top of this later coil.

Modern Day Tesla Coils

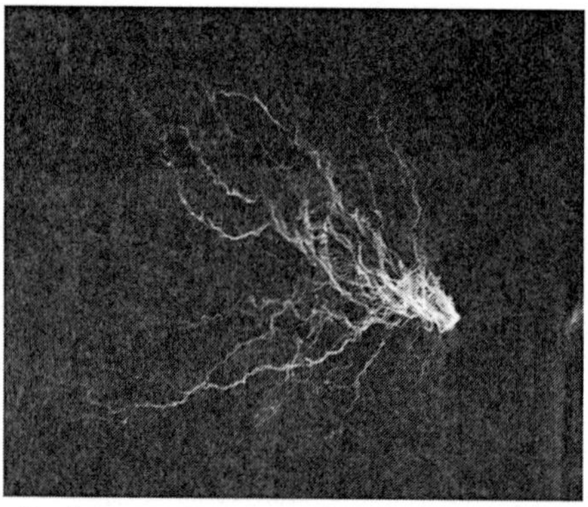

Figure 16: Electric discharge showing the lightning-like plasma filaments from a *Tesla coil.*

Modern high voltage enthusiasts usually build Tesla coils that are similar to some of Tesla's "later" air core designs. These typically consist of a primary tank circuit, a series LC (inductance-capacitance) circuit composed of a high voltage capacitor, spark gap and primary coil, and the secondary LC circuit, a series resonant circuit consisting of the secondary coil plus a terminal capacitance or "top load." In Tesla's more advanced design, the secondary LC circuit is composed of an air-core transformer secondary coil placed in series with a helical resonator. The helical coil is then connected to the terminal capacitance. Most modern coils use only a single helical coil comprising both the secondary and primary resonator. The terminal capacitance actually forms one 'plate' of a capacitor, the other 'plate' being the Earth (or "ground"). The primary LC circuit is tuned so that it resonates at the same frequency as the secondary LC circuit. The primary and secondary coils are magnetically coupled, creating a dual-tuned resonant air-core transformer. Earlier oil insulated Tesla coils needed large and long insulators at their high-voltage terminals to prevent discharge in air. Later version Tesla coils spread their electric fields over large distances to prevent high electrical stresses in the first place, thereby allowing operation in free air.

Most modern Tesla coils use simple toroids, typically fabricated from spun metal or flexible aluminium ducting, to control the high electrical field near the top of the secondary and to direct spark outward and away from the primary and secondary windings.

More advanced Tesla coil transmitters involve a more tightly coupled air core resonance transformer network or "master oscillator" the output of which is then fed another resonator, sometimes called the "extra coil." The principle is that energy accumulates in the extra coil and the role of transformer secondary is played by the separate master oscillator secondary; the roles are not shared by a single secondary. In some modern three-coil Magnifying transmitter systems the extra coil is placed some distance from the transformer. Direct magnetic coupling to the upper secondary is not desirable, since the third coil is designed to be driven by injecting RF current directly into the bottom end.

This particular Tesla coil configuration consists of a secondary coil in close inductive relation with a primary, and one end of which is connected to a ground-plate, while its other end is led through a separate self-induction coil (whose connection should always be made at, or near, the geometrical centre of that coil's circular aspect, in order to secure a symmetrical distribution of the current), and of a metallic cylinder carrying the current to the terminal. The primary coil may be excited by any desired source of high frequency current. The important requirement is that the primary and secondary sides must be tuned to the same resonant frequency to allow efficient transfer of energy between the primary and secondary resonant circuits. The conductor of the shaft to the terminal (topload) is in the form of a cylinder with smooth surface of a radius much larger than that of the spherical metal plates, and widens out at the bottom into a hood (which is slotted to avoid loss by eddy currents). The secondary coil is wound on a drum of insulating material, with its turns close together. When the effect of the small radius of curvature of the wire itself is overcome, the lower secondary coil behaves as a conductor of large radius of curvature, corresponding to that of the drum. The top of the extra coil may be extended up to the terminal U.S. Patent 1,119,732 and the bottom should be somewhat below the uppermost turn of the primary coil. This lessens the tendency of the charge to break out from the wire connecting both and to pass along the support.

Primary Switching

Modern day transistor or vacuum tube Tesla coils do not use a primary spark gap. Instead, the transistor(s) or vacuum tube(s) provide the switching or amplifying function necessary to generate RF power for the primary circuit. Solid-state Tesla coils use the lowest primary operating voltage, typically between 155 to 800 volts, and drive the primary winding using either a single, half-bridge, or full-bridge arrangement of bipolar transistors, MOSFETs or IGBTs to switch the primary current. Vacuum tube coils typically operate with plate voltages between 1500 and 6000 volts, while most spark gap coils operate with primary voltages of 6,000 to 25,000 volts. The primary winding of a traditional transistor Tesla coil is wound around only the bottom portion of the secondary (sometimes called the resonator). This helps to illustrate operation of the secondary as a pumped resonator. The primary *induces alternating* voltage into the bottom most portion of the secondary, providing regular "pushes" (similar to provided properly timed pushes to a playground swing). Additional energy is transferred from the primary to the secondary inductance and topload capacitance during each "push", and secondary output voltage builds (called *ring-up*). An electronic feedback circuit is usually used to adaptively synchronise the primary oscillator to the growing resonance in the secondary, and this is the only tuning consideration beyond the initial choice of a reasonable topload.

In a *dual resonant solid-state Tesla coil (DRSSTC)*, the electronic switching of the solid-state Tesla coil is combined with the resonant primary circuit of a spark-gap Tesla coil. The resonant primary circuit is formed by connecting a capacitor in series with the primary winding of the coil, so that the combination forms a seriestank circuit with a resonant frequency near that of the secondary circuit. Because of the additional resonant circuit, one manual and one adaptive tuning adjustment are necessary. Also, an interrupter is usually used to reduce the duty cycle of the switching bridge, in order to improve peak power capabilities; similarly, IGBTs are more popular in this application than bipolar transistors or MOSFETs, due to their superior power handling characteristics. Performance of a DRSSTC can be comparable to a medium power spark gap Tesla coil, and efficiency (as measured by spark length versus input power) can be significantly greater than a spark gap Tesla coil operating at the same input power.

Practical Aspects of Design

High Voltage Production: A large Tesla coil of more modern design often operates at very high peak power levels, up to many megawatts (millions of watts). It should therefore be adjusted and operated carefully, not only for efficiency and economy, but also for safety. If, due to improper tuning, the maximum voltage point occurs below the terminal, along the secondary coil, a discharge (spark) may break out and damage or destroy the coil wire, supports, or nearby objects.

Tesla experimented with these, and many other, circuit configurations. The Tesla coil primary winding, spark gap and tank capacitor are connected in series. In each circuit, the AC supply transformer charges the tank capacitor until its voltage is sufficient to break down the spark gap. The gap suddenly fires, allowing the charged tank capacitor to discharge into the primary winding. Once the gap fires, the electrical behaviour of either circuit is identical. Experiments have shown that neither circuit offers any marked performance advantage over the other.

However, in the typical circuit (above), the spark gap's short circuiting action prevents high frequency oscillations from 'backing up' into the supply transformer. In the alternate circuit, high amplitude high frequency oscillations that appear across the capacitor also are applied to the supply transformer's winding. This can induce corona discharges between turns that weaken and eventually destroy the transformer's insulation. Experienced Tesla coil builders almost exclusively use the top circuit, often augmenting it with low pass filters (resistor and capacitor (RC) networks) between the supply transformer and spark gap to help protect the supply transformer. This is especially important when using transformers with fragile high voltage windings, such as Neon-sign transformers (NSTs). Regardless of which configuration is used, the HV transformer must be of a type that self-limits its secondary current by means of internal leakage inductance. A normal (low leakage inductance) high voltage transformer must use an external limiter (sometimes called a ballast) to limit current. NSTs are designed to have high leakage inductance to limit their short circuit current to a safe level.

Tuning Precautions

The primary coil's resonant frequency should be tuned to that of the secondary, using low-power oscillations, then increasing the power until the apparatus has been brought under control. While tuning, a small projection (called a "breakout bump") is often added to the top terminal in order to stimulate corona and spark discharges (sometimes called streamers) into the surrounding air. Tuning can then be adjusted so as to achieve the longest streamers at a given power level, corresponding to a frequency match between the primary and secondary coil. Capacitive 'loading' by the streamers tends to lower the resonant frequency of a Tesla coil operating under full power. For a variety of technical reasons, toroids provide one of the most effective shapes for the top terminals of Tesla coils.

Air Discharges

While generating discharges, electrical energy from the secondary and toroid is transferred to the surrounding air as electrical charge, heat, light, and sound. The electric currents that flow through these discharges are actually due to the rapid shifting of quantities of charge from one place (the top terminal) to other places (nearby regions of air). The process is similar to charging or discharging a capacitor. The current that arises from shifting charges within a capacitor is called a displacement current. Tesla coil discharges are formed as a result of displacement currents as pulses of electrical charge are rapidly transferred between the high voltage toroid and nearby regions within the air (called space charge regions). Although the space charge regions around the toroid are invisible, they play a profound role in the appearance and location of Tesla coil discharges.

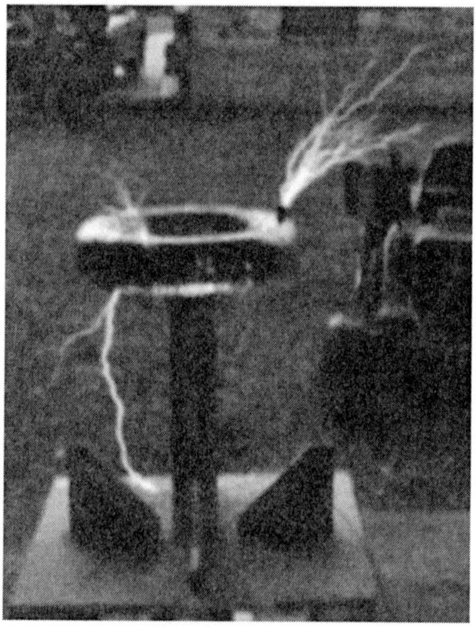

Figure 17: A small, later-type "*Tesla coil*" in operation. The output is giving 17-inch sparks. The diameter of the secondary is three inches. The power source is a 10000 V, 60 Hz current limited supply.

When the spark gap fires, the charged capacitor discharges into the primary winding, causing the primary circuit to oscillate. The oscillating primary current creates a magnetic field that couples to the secondary winding, transferring energy into the secondary side of the transformer and causing it to oscillate with the toroid capacitance. The energy transfer occurs over a number of cycles, and most of the energy that was originally in the primary side is transferred into the secondary side. The greater the magnetic coupling between windings, the shorter the time required to complete the energy transfer. As energy builds within the oscillating secondary circuit, the amplitude of the toroid's RF voltage rapidly increases, and the air surrounding the toroid begins to undergodielectric breakdown, forming a corona discharge.

As the secondary coil's energy (and output voltage) continue to increase, larger pulses of displacement current further ionise and heat the air at the point of initial breakdown. This forms a very conductive "root" of hotter plasma, called a leader, that projects outward from the toroid. The plasma within the leader is considerably hotter than a corona discharge, and is considerably more conductive. In fact, it has properties that are similar to an electric arc. The leader tapers and branches into thousands of thinner, cooler, hairlike discharges (called streamers).

The streamers look like a bluish 'haze' at the ends of the more luminous leaders, and it is the streamers that actually transfer charge between the leaders and toroid to nearby space charge regions. The displacement currents from countless streamers all feed into the leader, helping to keep it hot and electrically conductive.

The primary break rate of sparking Tesla coils is slow compared to the resonant frequency of the resonator-topload assembly. When the switch closes, energy is transferred from the primary LC circuit to the resonator where the voltage rings up over a short period of time up culminating in the electrical discharge. In a spark gap Tesla coil the primary-to-secondary energy transfer process happens repetitively at typical pulsing rates of 50–500 times per second, and previously formed leader channels don't get a chance to fully cool down between pulses. So, on successive pulses, newer discharges can build upon the hot pathways left by their predecessors.

This causes incremental growth of the leader from one pulse to the next, lengthening the entire discharge on each successive pulse. Repetitive pulsing causes the discharges to grow until the average energy that's available from the Tesla coil during each pulse balances the average energy being lost in the discharges (mostly as heat). At this point, dynamic equilibrium is reached, and the discharges have reached their maximum length for the Tesla coil's output power level. The unique combination of a rising high voltage Radio Frequency envelope and repetitive pulsing seem to be ideally suited to creating long, branching discharges that are considerably longer than would be otherwise expected by output voltage considerations alone. High voltage discharges create filamentary multi-branched discharges which are purplish blue in colour. High energy discharges create thicker discharges with fewer branches, are pale and luminous, almost white, and are much longer than low energy discharges, because of increased ionisation. There will be a strong smell of ozone and nitrogen oxides in the area. The important factors for maximum discharge length appear to be voltage, energy, and still air of low to moderate humidity. However, even more than 100 years later after the first use of Tesla coils, there are many aspects of Tesla coil discharges and the energy transfer process that are still not completely understood.

Applications

Tesla coil circuits were used commercially in sparkgap radio transmitters for wireless telegraphy until the 1920s, and in electrotherapy and pseudomedical devices such as violet ray. Today their main use is entertainment and educational displays. Tesla coils are built by many high-voltage enthusiasts, research institutions, science museums and independent experimenters. Although electronic circuit controllers have been developed, Tesla's original spark gap design is less expensive and has proven extremely reliable.

Wireless Transmission and Reception

The Tesla coil can also be used for wireless transmission. In addition to the positioning of the elevated terminal well above the top turn of the helical resonator, another difference from the sparking Tesla coil is the primary break rate. The optimised Tesla coil transmitter is a continuous wave oscillator with a break rate equalling the operating frequency. The combination of a helical resonator with an elevated terminal is also used for wireless reception. The Tesla coil receiver is intended for receiving the non-radiating electromagnetic field energy produced by the Tesla coil transmitter. The Tesla coil receiver is also adaptable for exploiting the ubiquitous vertical voltage gradient in the Earth's atmosphere. Tesla built and used various devices for detecting electromagnetic field energy. His early wireless apparatus operated on the basis of Hertzian waves or ordinary radio waves, electromagnetic waves that propagate in space without involvement of a conducting guiding surface. During his work at Colorado Springs, Tesla believed he had established electrical resonance of the entire Earth using the Tesla coil transmitter at his "Experimental Station."

Tesla stated one of the requirements of the World Wireless System was the construction of resonant receivers. The related concepts and methods are part of his wireless transmission system (US1119732 — Apparatus for Transmitting Electrical Energy — 1902 January 18).

Tesla made a proposal that there needed to be many more than thirty transmission-reception stations worldwide. In one form of receiving circuit the two input terminals are connected each to a mechanical pulse-width modulation device adapted to reverse polarity at predetermined intervals of time and charge a capacitor. This form of Tesla system receiver has means for commutating the current impulses in the charging circuit so as to render them suitable for charging the storage device, a device for closing the receiving-circuit, and means for causing the receiver to be operated by the energy accumulated.

A Tesla coil used as a receiver is referred to as a *Tesla receiving transformer*. The Tesla coil receiver acts as a step-down transformer with high current output. The parameters of a Tesla coil transmitter are identically applicable to it being a receiver (*e.g.*., an antenna circuit), due to reciprocity. Impedance, generally though, is not applied in an obvious way; for electrical impedance, the impedance at the load (*e.g.*., where the power is consumed) is most critical and, for a Tesla coil receiver, this is at the point of utilisation (such as at an induction motor) rather than at the receiving node. Complex impedance of an antenna is related to the electrical length of the antenna at the wavelength in use. Commonly, impedance is adjusted at the load with a tuner or a matching networks composed of inductors and capacitors.

A Tesla coil can receive electromagnetic impulses from atmospheric electricity and radiant energy, besides normal wireless transmissions. Radiant energy throws off with great velocity minute particles which are strongly electrified and other rays falling on the insulated-conductor connected to a condenser (i.e., a capacitor) can cause the condenser to indefinitely charge electrically. The helical resonator can be "shock excited" due to radiant energy disturbances not only at the fundamental wave at one-quarter wave-length but also is excited at its harmonics. Hertzian methods can be used to excite the Tesla coil receiver with limitations that result in great disadvantages for utilisation, though. The methods of ground conduction and the various induction methods can also be used to excite the Tesla coil receiver, but are again at a disadvantages for utilisation. The charging-circuit can be adapted to be energised by the action of various other disturbances and effects at a distance. Arbitrary and intermittent oscillations that are propagated via conduction to the receiving resonator will charge the receiver's capacitor and utilise the potential energy to greater effect. Various radiations can be used to charge and discharge conductors, with the radiations considered electromagnetic vibrations of various wavelengths and ionising potential. The Tesla receiver utilises the effects or disturbances to charge a storage device with energy from an external source (natural or man-made) and controls the charging of said device by the actions of the effects or disturbances (during succeeding intervals of time determined by means of such effects and disturbances corresponding in succession and duration of the effects and disturbances). The stored energy can also be used to operate the receiving device. The accumulated energy can, for example, operate a transformer by discharging through a primary circuit at predetermined times which, from the secondary currents, operate the receiving device.

While Tesla coils can be used for these purposes, much of the public and media attention is directed away from transmission-reception applications of the Tesla coil since electrical spark discharges are fascinating to many people. Regardless of this fact, Tesla did suggest that this variation of the Tesla coil could utilise the phantom loop effect to form a circuit to induct energy from the Earth's magnetic field and other radiant energy sources (including, but not limited to, electrostatics). With regard to Tesla's statements on the harnessing of natural phenomena to obtain electric power, he stated:

Ere many generations pass, our machinery will be driven by a power obtainable at any point of the universe. — "Experiments with Alternate Currents of High Potential and High Frequency" (February 1892)

Tesla stated that the output power from these devices, attained from Hertzian methods of charging, was low, but alternative charging means are available. Tesla receivers, operated correctly, act as a step-down transformer with high current output. There are, to date, no commercial power generation entities or businesses that have utilised this technology to full effect. The power levels achieved by Tesla coil receivers have, thus far, been a fraction of the output power of the transmitters.

High Frequency Electrical Safety

The 'Skin Effect' : The dangers of contact with high frequency electrical current are sometimes perceived as being less than at lower frequencies, because the subject usually doesn't feel pain or a 'shock'. This is often erroneously attributed to skin effect, a phenomenon that tends to inhibit alternating current from flowing inside conducting media. It was thought that in the body, Tesla currents travelled close to the skin surface, making them safer than lower frequency electric currents. In fact, in the early 1900s a major use of Tesla coils was to apply high frequency current directly to the body in electrotherapy. Although skin effect limits Tesla currents to the outer fraction of an inch in metal conductors, the 'skin depth' of human flesh at typical Tesla coil frequencies is still of the order of 60 inches (150 cm) or more. This means that high frequency currents will still preferentially flow through deeper, better conducting, portions of an experimenter's body such as the circulatory and nervous systems. The reason for the lack of pain is that a human being's nervous system does not sense the flow of potentially dangerous electrical currents above 15–20 kHz; essentially, in order for nerves to be activated, a significant number of ions must cross their membrane before the current (and hence voltage) reverses. Since the body no longer provides a warning 'shock', novices may touch the output streamers of small Tesla coils without feeling painful shocks. However, there is anecdotal evidence among Tesla coil experimenters that temporary tissue damage may still occur and be observed as muscle pain, joint pain, or tingling for hours or even days afterwards. This is believed to be caused by the damaging effects of internal current flow, and is especially common with continuous wave (CW), solid state or vacuum tube type Tesla coils. Some transformers can provide alternating current with such high frequencies that the skin depth becomes small enough for the voltage to be safe. Skin depth is inversely proportional to the root of the frequency, putting these frequencies in the megahertz range.

Large Tesla coils and magnifiers can deliver dangerous levels of high frequency current, and they can also develop significantly higher voltages (often 250,000–500,000 volts, or more). Because of the higher voltages, large systems can deliver higher energy, potentially lethal, repetitive high voltage capacitor discharges from their top terminals. Doubling the output voltage quadruples the electrostatic energy stored in a given top terminal capacitance. If an unwary experimenter accidentally places himself in path of the high voltage capacitor discharge to ground, the low current electric shock can cause involuntary spasms of major muscle groups and may induce life-threatening ventricular fibrillation and cardiac arrest. Even lower power vacuum tube or solid state Tesla coils can deliver RF currents that are capable of causing temporary internal tissue, nerve, or joint damage through Joule heating. In addition, an RF arc can carbonise flesh, causing a painful and dangerous bone-deep RF burn that may take months to heal. Because of these risks, knowledgeable experimenters avoid contact with streamers from all but the smallest systems. Professionals usually use other means of protection such as a Faraday cage or a chain mail suit to prevent dangerous currents from entering their body.

The most serious dangers associated with Tesla coil operation are associated with the primary circuit. It is the primary circuit that is capable of delivering a sufficient current at a significant voltage to stop the heart of a careless experimenter. Because these components are not the source of the trademark visual or auditory coil effects, they may easily be overlooked as the chief source of hazard. Should a high frequency arc strike the exposed primary coil while, at the same time, another arc has also been allowed to strike to a person, the ionised gas of the two arcs forms a circuit that may conduct lethal, low-frequency current from the primary into the person.

Further, great care should be taken when working on the primary section of a coil even when it has been disconnected from its power source for some time. The tank capacitors can remain charged for days with enough energy to deliver a fatal shock. Proper designs should always include 'bleeder resistors' to bleed off stored charge from the capacitors. In addition, a safety shorting operation should be performed on each capacitor before any internal work is performed.

Instances and Devices

Magnifier Configurations:

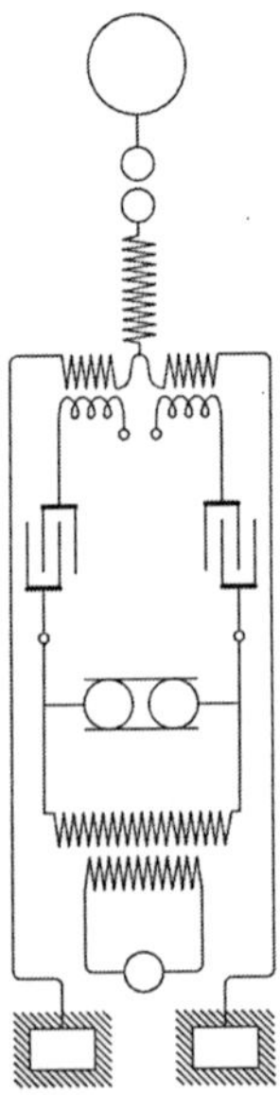

Figure 18: Classically driven configuration.

Figure 19: Later-type driven configuration. Pancake may be horizontal; lead to resonator is kept clear of it.

Tesla's Colorado Springs laboratory possessed one of the largest Tesla coils ever built, known as the "Magnifying Transmitter". The Magnifying Transmitter is somewhat different from classic 2-coil Tesla coils. A Magnifier uses a 2-coil 'driver' to excite the base of a third coil ('resonator') that is located some distance from the driver. The operating principles of both systems are similar. The world's largest currently existing 2-coil Tesla coil is a 130,000-watt unit, part of a 38-foot-tall (12 m) sculpture. It is owned by Alan Gibbs and currently resides in a private sculpture park at Kakanui Point near Auckland, New Zealand.

The Tesla coil is an early predecessor (along with the induction coil) of a more modern device called a flyback transformer, which provides the voltage needed to power the cathode ray tube used in some televisions and computer monitors. The disruptive discharge coil remains in common use as the *ignition coil* or *spark coil* in the ignition system of an internal combustion engine. These two devices do not use resonance to accumulate energy, however, which is the distinguishing feature of a Tesla coil. They do use inductive "kick", the forced, abrupt decay of the magnetic field, such that a voltage is provided by the coil at its primary terminals that is much greater than the voltage that was applied to establish the magnetic field, and it is this higher voltage that is then multiplied by the transformer turns ratio. Thus, they do store energy, and a Tesla resonator stores energy. A modern, low power variant of the Tesla coil is also used to power plasma globe sculptures and similar devices.

Scientists working with a glass vacuum line (e.g. chemists working with volatile substances in the gas phase, inside a system of glass tubes, taps and bulbs) test for the presence of tiny pin-holes in the apparatus (especially a newly blown piece of glassware) using high voltage discharges, such as a Tesla coil produces. When the system is evacuated and the discharging end of the coil moved over the glass, the discharge travels through any pin-hole immediately below it and thus illuminates the hole, indicating points that need to be annealed or re-blown before they can be used in an experiment.

Popularity

Tesla coils are very popular devices among certain electrical engineers and electronics enthusiasts. Builders of Tesla coils as a hobby are called "coilers". A very large tesla coil, designed and built by Syd Klinge, is shown every year at the Coachella Valley Music and Arts Festival, in Coachella, Indio, California, USA. There are "coiling" conventions where people attend with their home-made Tesla coils and other electrical devices of interest.

Low power Tesla coils are also sometimes used as a high voltage source for Kirlian photography.

Tesla coils can also be used to create music by modulating the system's effective "break rate" (i.e., the rate and duration of high power RF bursts) via MIDI data and a control unit. The actual MIDI data is interpreted by a microcontroller which converts the MIDI data into a PWM output which can be sent to the Tesla coil via a fibre optic interface.

Chapter 4

Inductively Coupled Plasma

An inductively coupled plasma (ICP) is a type of plasmasource in which the energy is supplied by electric currents which are produced by electromagnetic induction, that is, by time-varying magnetic fields.

Operation

There are two types of ICP geometries: planar and cylindrical. In planar geometry, the electrode is a coil of flat metal wound like a spiral. In cylindrical geometry, it is like a helical spring.

When a time-varying electric current is passed through the coil, it creates a time-varying magnetic field around it, which in turn induces azimuthal electric currents in the rarefied gas, leading to the formation of a plasma. Argon is one example of a commonly-used rarefied gas.

Applications

Plasma temperatures can range between 6 000 K and 10 000 K, comparable to the surface of the sun. ICP discharges are of relatively high electron density, on the order of 10^{15} cm^{-3}. As a result, ICP discharges have wide applications where a high-density plasma is needed.

- ICP-AES, a type of atomic emission spectrometry
- ICP-MS, a type of mass spectrometry.
- ICP-RIE, a type of reactive-ion etching.

Another benefit of ICP discharges is that they are relatively free of contamination because the electrodes are completely outside the reaction chamber. By contrast, in a capacitively coupled plasma (CCP), the electrodes are often placed inside the reactor and are thus exposed to the plasma and subsequent reactive chemical species.

Lightning

Lightning is an atmospheric electrostatic discharge (spark) accompanied by thunder, which typically occurs during thunderstorms, and sometimes during volcanic eruptions or dust storms. From this discharge of atmospheric electricity, a leader of a bolt of lightning can travel at speeds of 220,000 km/h (140,000 mph), and can reach temperatures approaching 30,000 °C (54,000 °F), hot enough to fuse silica sand into glass channels known as fulgurites, which are normally hollow and can extend some distance into the ground. There are some 16 million lightning storms in the world every year. Lightning causes ionisation in the air through which it travels, leading to the formation of nitric oxide and ultimately, nitric acid, of benefit to plant life below.

Lightning can also occur within the ash clouds from volcanic eruptions, or can be caused by violent forest fires which generate sufficient dust to create a static charge.

How lightning initially forms is still a matter of debate. Scientists have studied root causes ranging from atmospheric perturbations (wind, humidity, friction, and atmospheric pressure) to the impact of solar wind and accumulation of charged solar particles. Ice inside a cloud is thought to be a key element in lightning development, and may cause a forcible separation of positive and negative charges within the cloud, thus assisting in the formation of lightning.

The irrational fear of lightning (and thunder) is astraphobia. The study or science of lightning is called fulminology, and someone who studies lightning is referred to as a fulminologist.

History of Lightning Research

Benjamin Franklin (1706–1790) endeavoured to test the theory that sparks shared some similarity with lightning by using a spire which was being erected in Philadelphia, United States. While waiting for completion of the spire, he got the idea to use a flying object such as a kite. During the next thunderstorm, which was in June 1752, it was reported that he raised a kite. He was accompanied by his son as an assistant. On his end of the string he attached a key, and he tied it to a post with a silk thread. As time passed, Franklin noticed the loose fibres on the string stretching out; he then brought his hand close to the key and a spark jumped the gap. The rain which had fallen during the storm had soaked the line and made it conductive.

Franklin was not the first to perform the kite experiment. Thomas-François Dalibard and De Lors conducted it at Marly-la-Ville in France, a few weeks before Franklin's experiment. In his autobiography (written 1771–1788, first published 1790), Franklin clearly states that he performed this experiment after those in France, which occurred weeks before his own experiment, without his prior knowledge as of 1752.

Figure 1: Lightning strikes the Eiffel Tower, Francein 1902.

As news of the experiment and its particulars spread, others attempted to replicate it. However, experiments involving lightning are always risky and frequently fatal. One of the most well-known deaths during the spate of Franklin imitators was that of Professor Georg Richmann of Saint Petersburg, Russia. He created a set-up similar to Franklin's, and was attending a meeting of the Academy of Sciences when he heard thunder. He ran home with his engraver to capture the event for posterity. According to reports, while the experiment was under way, ball lightning appeared and collided with Richmann's head, killing him.

Although experiments from the time of Benjamin Franklin showed that lightning was a discharge of static electricity, there was little improvement in theoretical understanding of lightning (in particular how it was generated) for more than 150 years. The impetus for new research came from the field of power engineering: as power transmission lines came into service, engineers needed to know much more about lightning in order to adequately protect lines and equipment. In 1900, Nikola Teslagenerated artificial lightning by using a large Tesla coil, enabling the generation of enormously high voltages sufficient to create lightning.

Properties

Lightning can occur with both positive and negative polarity. An average bolt of negative lightning carries an electric current of 30,000 amperes (30 kA), and transfers fifteen coulombs of electric chargeand 500 megajoules of energy. Large bolts of lightning can carry up to 120 kA and 350 coulombs. An average bolt of positive lightning carries an electric current of about 300 kA — about 10 times that of negative lightning.

The voltage involved for both isproportional to the length of the bolt. However, lightning leader development is not just a matter of the electrical breakdown of air, which occurs at a voltage gradient of about 1 megavolts per metre (MV/m). The ambient electric fieldsrequired for lightning leader propagation can be one or two orders of magnitude (10^{-2}) less than the electrical breakdown strength. The potential ("voltage") gradient inside a well-developed return-stroke channel is on the order of hundreds of volts per metre (V/m) due to intense channel ionisation, resulting in a true power output on the order of one megawatt per meter (MW/m) for a vigorous return stroke current of 100 kA. The average *peak* power output of a single lightning stroke is about one trillion watts — one terawatt (10^{12} W), and the stroke lasts for about 30 millionths of a second — 30microseconds.

Lightning rapidly heats the air in its immediate vicinity to about 20,000 °C (36,000 °F) — about three times the temperature of the surface of the Sun. The sudden heating effect and the expansion of heated air gives rise to a supersonic shock wave in the surrounding clear air. It is this shock wave, once it decays to an acoustic wave, that is heard as thunder.

The return stroke of a lightning bolt follows a charge channel about a centimetre (0.4 in) wide.

Different locations have different potentials and currents for an average lightning strike. In the United States, for example, Florida experiences the largest number of recorded strikes in a given period during the summer season , has very sandy soils in some areas, and electrically conductive water-saturated soils in others. As much of Florida lies on a peninsula, it is bordered by the ocean on three sides. The result is the daily development of sea and lake breeze boundaries that collide and produce thunderstorms.

NASA scientists have found that electromagnetic radiation created by lightning in clouds only a few miles high can create a safe zone in the Van Allen radiation belts that surround the earth. This zone, known as the "Van Allen Belt slot", may be a safe haven for satellites in middle Earth orbits (MEOs), protecting them from the Sun's intense radiation.

Formation

Positive lightning (a rarer form of lightning that originates from positively charged regions of the thundercloud) does not generally fit the preceding pattern.

Cloud Particle Collision Hypothesis

According to this cloud particle charging hypothesis, charges are separated when ice crystals rebound off graupel. Charge separation appears to require strong updrafts which carry water droplets upward, supercooling them to between -10 and -40 °C. These water droplets collide with ice crystals to form a soft ice-water mixture called graupel. Collisions between ice crystals and graupel pellets usually results in positive charge being transferred to the ice crystals, and negative charge to the graupel. Updrafts drive the less heavy ice crystals upwards, causing the cloud top to accumulate increasing positive charge. Gravity causes the heavier negatively charged graupel to fall toward the middle and lower portions of the cloud, building up an increasing negative charge. Charge separation and accumulation continue until the electrical potential becomes sufficient to initiate a lightning discharge, which occurs when the distribution of positive and negative charges forms a sufficiently strong electric field.

Polarisation Mechanism Hypothesis

The mechanism by which charge separation happens is still the subject of research. Another hypothesis is the polarisation mechanism, which has two components:

1. Falling droplets of ice and rain become electrically polarised as they fall through the Earth's natural electric field;
2. Colliding/rebounding cloud particles become oppositely charged.

There are several hypotheses for the origin of charge separation.

Lightning Initiation

Even assuming an electric field has been established, the mechanism by which the lightning discharge begins is not well known. Electric field measurements in thunderclouds are typically not large enough to directly initiate a discharge. Many hypotheses have been proposed, ranging from including runaway breakdown to locally enhanced electric fields near elongated water droplets or ice crystals. Percolation theory, especially for the case of biased percolation, describe random connectivity phenomena, which produce an evolution of connected structures similar to that of lightning strikes.

Leader Formation and the Return Stroke

As a thundercloud moves over the surface of the Earth, an electric charge equal to but opposite the charge of the base of the thundercloud is induced in the Earth below the cloud. The induced ground charge follows the movement of the cloud, remaining underneath it.

An initial bipolar discharge, or path of ionised air, starts from a negatively charged region of mixed water and ice in the thundercloud. Discharge ionised channels are known as leaders. The positive and negative charged leaders, generally a "stepped leader", proceed in opposite directions. The negatively-charged one proceeds downward in a number of quick jumps (steps). 90 percent of the leaders exceed 45 m (148 ft) in length, with most in the order of 50 to 100 m (164 to 492 feet). As it continues to descend, the stepped leader may branch into a number of paths. The progression of stepped leaders takes a comparatively long time (hundreds of milliseconds) to approach the ground. This initial phase involves a relatively small electric current (tens or hundreds of amperes), and the leader is almost invisible when compared with the subsequent lightning channel.

When a stepped leader approaches the ground, the presence of opposite charges on the ground enhances the strength of the electric field. The electric field is strongest on ground-connected objects whose tops are closest to the base of the thundercloud, such as trees and tall buildings. If the electric field is strong enough, a conductive discharge (called a positive streamer) can develop from these points. This was first theorised by Heinz Kasemir. As the field increases, the positive streamer may evolve into a hotter, higher current leader which eventually connects to the descending stepped leader from the cloud. It is also possible for many streamers to develop from many different objects simultaneously, with only one connecting with the leader and forming the main discharge path. Photographs have been taken on which non-connected streamers are clearly visible.

Once a channel of ionised air is established between the cloud and ground this becomes a *path of least resistance* and allows for a much greater current to propagate from the Earth back up the leader into the cloud. This is the *return stroke* and it is the most luminous and noticeable part of the lightning discharge.

Discharge

When the electric field becomes strong enough, an electrical discharge (the bolt of lightning) occurs within clouds or between clouds and the ground. During the strike, successive portions of air become a conductive discharge channel as the electrons and positive ions of air molecules are pulled away from each other and forced to flow in opposite directions.

The electrical discharge rapidly superheats the discharge channel, causing the air to expand rapidly and produce a shock wave heard as thunder. The rolling and gradually dissipating rumble of thunder is caused by the time delay of sound coming from different portions of a long stroke.

Re-strike

High speed videos (examined frame-by-frame) show that most lightning strikes are made up of multiple individual strokes. A typical strike is made of 3 or 4 strokes, though there may be more.

Each re-strike is separated by a relatively large amount of time, typically 40 to 50 milliseconds. Re-strikes can cause a noticeable "strobe light" effect.

Each successive stroke is preceded by intermediate dart leader strokes akin to, but weaker than, the initial stepped leader. The stroke usually re-uses the discharge channel taken by the previous stroke. The variations in successive discharges are the result of smaller regions of charge within the cloud being depleted by successive strokes.

The sound of thunder from a lightning strike is prolonged by successive strokes.

Types of Lighting

Some lightning strikes exhibit particular characteristics; scientists and the general public have given names to these various types of lightning. The lightning that is most-commonly observed is streak lightning. This is nothing more than the return stroke, the visible part of the lightning stroke. The majority of strokes occur inside a cloud so we do not see most of the individual return strokes during a thunderstorm.

Cloud-to-Ground Lightning

This is the best known and second most common type of lightning. Of all the different types of lightning, it poses the greatest threat to life and property since it strikes the ground. Cloud-to-ground (CG) lightning is a lightning discharge between a cumulonimbus cloud and the ground. It is initiated by a leader stroke moving down from the cloud.

Figure 2: Cloud-to-ground lightning

Bead Lightning

Bead lightning is a type of cloud-to-ground lightning which appears to break up into a string of short, bright sections, which last longer than the usual discharge channel. It is relatively rare. Several theories have been proposed to explain it; one is that the observer sees portions of the lightning channel end on, and that these portions appear especially bright. Another is that, in bead lightning, the width of the lightning channel varies; as the lightning channel cools and fades, the wider sections cool more slowly and remain visible longer, appearing as a *string of beads*.

Ribbon Lightning

Ribbon lightning occurs in thunderstorms with high cross winds and multiple return strokes. The wind will blow each successive return stroke slightly to one side of the previous return stroke, causing a ribbon effect.

Staccato Lightning

Staccato lightning is a cloud-to-ground lightning (CG) strike which is a short-duration stroke that (often but not always) appears as a single very bright flash and often has considerable branching. These are often found in the visual vault area near the mesocyclone of rotating thunderstorms and coincides with intensification of thunderstorm updrafts. A similar cloud-to-cloud strike consisting of a brief flash over a small area, appearing like a blip, also occurs in a similar area of rotating updrafts.

Forked Lightning

Forked lightning is a name, not in formal usage, for cloud-to-ground lightning that exhibits branching of its path.

Ground-to-Cloud Lightning

Ground-to-cloud lightning is a lightning discharge between the ground and a cumulonimbus cloud initiated by an upward-moving leader stroke. This type of lightning forms when negatively charged ions called the stepped leader rise up from the ground and meet the positively charged ions in a cumulonimbus cloud. Then the strike goes back to the ground as the return stroke. This is also called positive lightning.

Cloud-to-Cloud Lightning

Figure 3: Multiple paths of cloud-to-cloud lightning, Swifts Creek, Australia

Lightning discharges may occur between areas of cloud without contacting the ground. When it occurs between two separate clouds it is known as *inter-cloud lightning*, and when it occurs between areas of differing electric potential within a single cloud it is known as intra-*cloud lightning*. Intra-cloud lightning is the most frequently occurring type.

These are most common between the upper anvil portion and lower reaches of a given thunderstorm. This lightning can sometimes be observed at great distances at night as so-called "heat lightning". In such instances, the observer may see only a flash of light without hearing any thunder. The "heat" portion of the term is a folk association between locally experienced warmth and the distant lightning flashes.

Another terminology used for cloud–cloud or cloud–cloud–ground lightning is "Anvil Crawler", due to the habit of the charge typically originating from beneath or within the anvil and scrambling through the upper cloud layers of a thunderstorm, normally generating multiple branch strokes which are dramatic to witnesses. These are usually seen as a thunderstorm passes over the observer or begins to decay. The most vivid crawler behaviour occurs in well developed thunderstorms that feature extensive rear anvil shearing.

Sheet Lightning

Sheet lightning is an informal name for cloud-to-cloud lightning that exhibits a diffuse brightening of the surface of a cloud, caused by the actual discharge path being hidden. The lightning itself cannot be seen by the spectator, so it appears as only a flash, or a sheet of light.

Heat Lightning

Heat lightning is a common name for a lightning flash that appears to produce no thunder because it occurs too far away for the thunder to be heard. The sound waves dissipate before they reach the observer.

Dry Lightning

Figure 4: Volcanic material thrust high into the atmosphere can trigger lightning.

Dry lightning is a term in Canada and the United States for lightning that occurs with no precipitation at the surface. This type of lightning is the most common natural cause of wildfires. Pyrocumulus clouds produce lightning for the same reason that it is produced by cumulonimbus clouds. When the higher levels of the atmosphere are cooler, and the surface is warmed to extreme temperatures due to a wildfire, volcano, etc., convection will occur, and the convection produces lightning. Therefore, fire can beget dry lightning through the development of more dry thunderstorms which cause more fires.

Rocket Lightning

It is a form of cloud discharge, generally horizontal and at cloud base, with a luminous channel appearing to advance through the air with visually resolvable speed, often intermittently.

Positive Lightning

Unlike the far more common "negative" lightning, positive lightning occurs when a positive charge is carried by the top of the clouds (generally anvil clouds) rather than the ground. Generally, this causes the leader arc to form in the anvil of the cumulonimbus and travel horizontally for several miles before veering down to meet the negatively charged streamer rising from the ground. The bolt can strike anywhere within several miles of the anvil of the thunderstorm, often in areas experiencing clear or only slightly cloudy skies; they are also known as "bolts from the blue" for this reason. Positive lightning makes up less than 5% of all lightning strikes. Because of the much greater distance they must travel before discharging, positive lightning strikes typically carry six to ten times the charge and voltage difference of a negative bolt and last around ten times longer. During a positive lightning strike, huge quantities of ELF and VLF radio waves are generated.

As a result of their greater power, as well as lack of warning, positive lightning strikes are considerably more dangerous. At the present time, aircraft are not designed to withstand such strikes, since their existence was unknown at the time standards were set, and the dangers unappreciated until the destruction of a glider in 1999. The standard in force at the time of the crash, Advisory Circular AC 20-53A, was replaced by Advisory Circular AC 20-53B in 2006, however it is unclear whether adequate protection against positive lighting was incorporated.

Positive lightning is also now believed to have been responsible for the 1963 in-flight explosion and subsequent crash of Pan Am Flight 214, a Boeing 707. Due to the dangers of lightning, aircraft operating in U.S. airspace have been required to have lightning discharge wicks to reduce the damage by a lightning strike, but these measures may be insufficient for positive lightning.

Positive lightning has also been shown to trigger the occurrence of upper atmosphere lightning. It tends to occur more frequently in winter storms, as with thunder snow, and at the end of a thunderstorm.

Ball Lightning

Ball lightning may be an atmospheric electrical phenomenon, the physical nature of which is still controversial. The term refers to reports of luminous, usually spherical objects which vary from pea-sized to several metres in diameter. It is sometimes associated with thunderstorms, but unlike lightning flashes, which last only a fraction of a second, ball lightning reportedly lasts many seconds. Ball lightning has been described by eyewitnesses but rarely recorded by meteorologists. Scientific data on natural ball lightning is scarce owing to its infrequency and unpredictability. The presumption of its existence is based on reported public sightings, and has therefore produced somewhat inconsistent findings.

Laboratory experiments have produced effects that are visually similar to reports of ball lightning, but at present, it is unknown whether these are actually related to any naturally occurring phenomenon. One theory is that ball lightning may be created when lightning strikes silicon in soil, a phenomenon which has been duplicated in laboratory testing. Given inconsistencies and the lack of reliable data and completely contradicting and unpredictable behaviour, the true nature of ball lightning is still unknown and was often regarded as a fantasy or a hoax. Reports of the phenomenon were dismissed for lack of physical evidence, and were often regarded the same way as UFO sightings. Severely contradicting descriptions of ball lightning makes it impossible even to create plausible hypothesis that will take into account described behaviour.

One theory that may account for this wider spectrum of observational evidence is the idea of combustion inside the low-velocity region of spherical vortex breakdown of a natural vortex (e.g., the 'Hill's spherical vortex'). Natural ball lightning appears infrequently and unpredictably, and is therefore rarely (if ever truly) photographed. However, several purported photos and videos exist. Perhaps the most famous story of ball lightning unfolded when 18th-century physicist Georg Wilhelm Richmann installed a lightning rod in his home and was struck in the head - and killed - by a "pale blue ball of fire."

Upper-Atmospheric Lightning

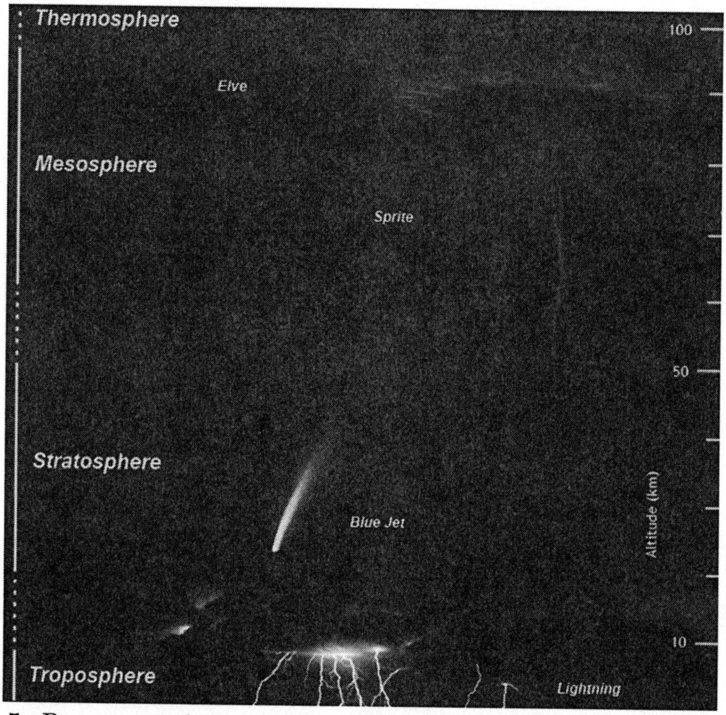

Figure 5: Representation of upper-atmospheric lightning and electrical-discharge phenomena

Reports by scientists of strange lightning phenomena about storms date back to at least 1886. However, it is only in recent years that fuller investigations have been made. This has sometimes been called megalightning.

Sprites

Sprites are large-scale electrical discharges that occur high above a thunderstorm cloud, or cumulonimbus, giving rise to a quite varied range of visual shapes. They are triggered by the discharges of positive lightning between the thundercloud and the ground. The phenomena were named after the mischievous sprite (air spirit) Puck in Shakespeare's *A Midsummer Night's Dream*. They normally are coloured reddish-orange or greenish-blue, with hanging tendrils below and arcing branches above their location, and can be preceded by a reddish halo. They often occur in clusters, lying 50 kilometres (31 mi) to 90 kilometres (56 mi) above the Earth's surface. Sprites were first photographed on July 6, 1989 by scientists from the University of Minnesota and have since been witnessed tens of thousands of times. Sprites have been mentioned as a possible cause in otherwise unexplained accidents involving high altitude vehicular operations above thunderstorms.

Blue Jets

Blue jets differ from sprites in that they project from the top of the cumulonimbus above a thunderstorm, typically in a narrow cone, to the lowest levels of the ionosphere 25 miles (40 km) to 50 miles (80 km) above the earth. They are also brighter than sprites and, as implied by their name, are blue in colour. They were first recorded on October 21, 1989, on a video taken from the space shuttle as it passed over Australia, and subsequently extensively documented in 1994 during aircraft research flights by the University of Alaska.

On September 14, 2001, scientists at the Arecibo Observatory photographed a huge jet double the height of those previously observed, reaching around 50 miles (80 km) into the atmosphere. The jet was located above a thunderstorm over the ocean, and lasted under a second. Lightning was initially observed travelling up at around 50,000 m/s in a similar way to a typical *blue jet*, but then divided in two and sped at 250,000 m/s to the ionosphere, where they spread out in a bright burst of light. On July 22, 2002, five gigantic jets between 60 and 70 km (35 to 45 miles) in length were observed over the South China Sea from Taiwan, reported in *Nature*. The jets lasted under a second, with shapes likened by the researchers to giant trees and carrots.

Elves

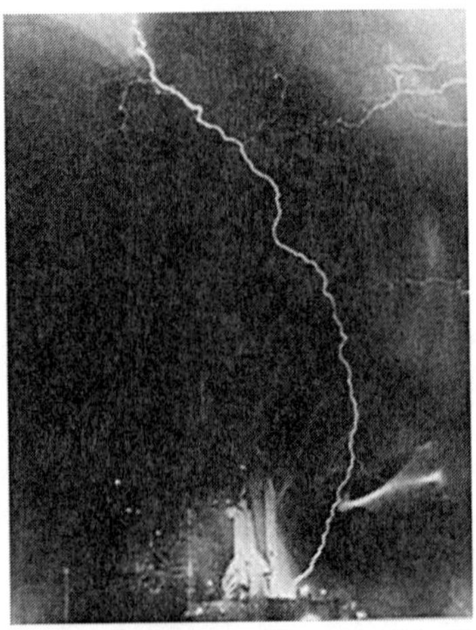

Figure 6: Lightning strikes the Space Shuttle *Challenger* before the launch of STS-8.

Elves often appear as dim, flattened, circular in the horizontal plane, expanding glows around 250 miles (400 km) in diameter that last for, typically, just one millisecond. They occur in the ionosphere 60 miles (97 km) above the ground over thunderstorms. Their colour was a puzzle for some time, but is now believed to be a red hue. Elves were first recorded on another shuttle mission, this time recorded off French Guiana on October 7, 1990. Elves is an acronym for Emissions of Light and Very Low Frequency Perturbations from Electromagnetic Pulse Sources. This refers to the process by which the light is generated; the excitation of nitrogen molecules due to electron collisions (the electrons possibly having been energised by the electromagnetic pulse caused by a discharge from the Ionosphere).

Triggered Lightning

Rocket-Triggered: Lightning has been triggered by launching lightning rockets carrying spools of wire into thunderstorms. The wire unwinds as the rocket ascends, providing a path for lightning. These bolts are typically very straight due to the path created by the wire.

Lightning has also been triggered directly by other human activities: Flying aircraft can trigger lightning. Furthermore, lightning struck Apollo 12 soon after takeoff, and has struck soon after thermonuclear explosions.

Volcanically Triggered

There are three types of volcanic lightning:

- Extremely large volcanic eruptions, which eject gases and material high into the atmosphere, can trigger lightning. This phenomenon was documented by Pliny The Elder during the 79 AD eruption of Vesuvius, in which he perished.
- An intermediate type which comes from a volcano's vents, sometimes 1.8 miles (2.9 km) long.
- Small spark-type lightning about 3 feet (0.91 m) long lasting a few milliseconds.

Laser-Triggered

Since the 1970s, researchers have attempted to trigger lightning strikes by means of infrared or ultraviolet lasers, which create a channel of ionised gas through which the lightning would be conducted to ground. Such triggering of lightning is intended to protect rocket launching pads, electric power facilities, and other sensitive targets.

In New Mexico, U.S., scientists tested a new terawatt laser which provoked lightning. Scientists fired ultra-fast pulses from an extremely powerful laser thus sending several terawatts into the clouds to call down electrical discharges in storm clouds over the region. The laser beams sent from the laser make channels of ionised molecules known as "filaments". Before the lightning strikes earth, the filaments lead electricity through the clouds, playing the role of lightning rods. Researchers generated filaments that lived too short a period to trigger a real lightning strike. Nevertheless, a boost in electrical activity within the clouds was registered. According to the French and German scientists, who ran the experiment, the fast pulses sent from the laser will be able to provoke lightning strikes on demand. Statistical analysis showed that their laser pulses indeed enhanced the electrical activity in the thundercloud where it was aimed—in effect they generated small local discharges located at the position of the plasma channels.

Extraterrestrial Lightning

Lightning requires the electrical breakdown of a gas, so it cannot exist in a visual form in the vacuum of space. However, lightning has been observed within the atmospheres of other planets, such as Venus, Jupiter and Saturn. Lightning on Venus is still a controversial subject after decades of study. During the Soviet Venera and U.S. Pioneer missions of the 1970s and '80s, signals suggesting lightning may be present in the upper atmosphere were detected. However, recently the Cassini–Huygens mission fly-by of Venus detected no signs of lightning at all. Despite this, it has been suggested that radio pulses recorded by the spacecraft Venus Express may originate from lightning on Venus.

High Energy Radiation Emissions Due to Lightning

The production of X-rays by a bolt of lightning was theoretically predicted as early as 1925 but no evidence was found until 2001/2002, when researchers at the New Mexico Institute of Mining and Technology detected X-ray emissions from an induced lightning strike along a wire trailed behind a rocket shot into a storm cloud. In the same year University of Florida and Florida Tech researchers used an array of electric field and X-ray detectors at a lightning research facility in North Florida to confirm that natural lightning makes X-rays in large quantities. The cause of the X-ray emissions is still a matter for research, as the temperature of lightning is too low to account for the X-rays observed.

Terrestrial Gamma-Ray Flashes

A number of observations by space-based telescopes have revealed even higher energy gamma ray emissions, the so-called terrestrial gamma-ray flashes (TGFs). These observations pose a challenge to current theories of lightning, especially with the discovery of the clear signatures of antimatter produced in lightning.

Figure 7: Double lightning

It has been discovered in the past 15 years that among the processes of lightning is some mechanism capable of generating gamma rays, which escape the atmosphere and are observed by orbiting spacecraft. Brought to light by NASA's Gerald Fishman in 1994 in an article in *Science*, these so-called terrestrial gamma-ray flashes (TGFs) were observed by accident, while he was documenting instances of extraterrestrial gamma ray bursts observed by the Compton Gamma Ray Observatory (CGRO). TGFs are much shorter in duration, however, lasting only about 1 ms.

Professor Umran Inan of Stanford University linked a TGF to an individual lightning stroke occurring within 1.5 ms of the TGF event, proving for the first time that the TGF was of atmospheric origin and associated with lightning strikes.

CGRO recorded only about 77 events in 10 years; however, more recently the Reuven Ramaty High Energy Solar Spectroscopic Imager (RHESSI) spacecraft, as reported by David Smith of UC Santa Cruz, has been observing TGFs at a much higher rate, indicating that these occur about 50 times per day globally (still a very small fraction of the total lightning on the planet). The energy levels recorded exceed 20 MeV.

Scientists from Duke University have also been studying the link between certain lightning events and the mysterious gamma ray emissions that emanate from the Earth's own atmosphere, in light of newer observations of TGFs made by RHESSI. Their study suggests that this gamma radiation fountains upward from starting points at surprisingly low altitudes in thunderclouds.

Steven Cummer, from Duke University's Pratt School of Engineering, said, "These are higher energy gamma rays than come from the sun. And yet here they are coming from the kind of terrestrial thunderstorm that see here all the time."

Early hypotheses of this pointed to lightning generating high electric fields and driving relativistic runaway electron avalanche at altitudes well above the cloud where the thin atmosphere allows gamma rays to easily escape into space, similar to the way sprites are generated. Subsequent evidence however, has suggested instead that TGFs may be produced by driving relativistic electron avalanches within or just above high thunderclouds. Though hindered by atmospheric absorption of the escaping gamma rays, these theories do not require the exceptionally intense lightning that high altitude theories of TGF generation rely on.

The role of TGFs and their relationship to lightning remains a subject of ongoing scientific study.

Sound

Because the electrostatic discharge of terrestrial lightning superheats the air to plasma temperatures along the length of the discharge channel in a short duration, kinetic theory dictates gaseous molecules undergo a rapid increase in pressure and thus expand outward from the lightning creating a shock wave audible as thunder. Since the sound waves propagate not from a single point source but along the length of the lightning's path, the sound origin's varying distances from the observer can generate a rolling or rumbling effect. Perception of the sonic characteristics is further complicated by factors such as the irregular and possibly branching geometry of the lightning channel, by acoustic echoing from terrain, and by the typically multiple-stroke characteristic of the lightning strike.

Since light travels at a significantly greater speed than ound through air, an observer can approximate the distance to the strike by timing the interval between the visible lightning and the audible thunder it generates. At standard atmospheric temperature and pressures near ground level, sound will travel at roughly 343 m/s (1125 ft/sec); a lightning flash preceding its thunder by five seconds would be about one mile (1.6 km) distant. A flash preceding thunder by three seconds is about one kilometre distant. Consequently, a lightning strike observed at a very close distance (within 100 meters) will be accompanied by the sound of a loud snap, thunder almost instantaneously and the smell of ozone(O_3).

Lightning-Induced Magnetism

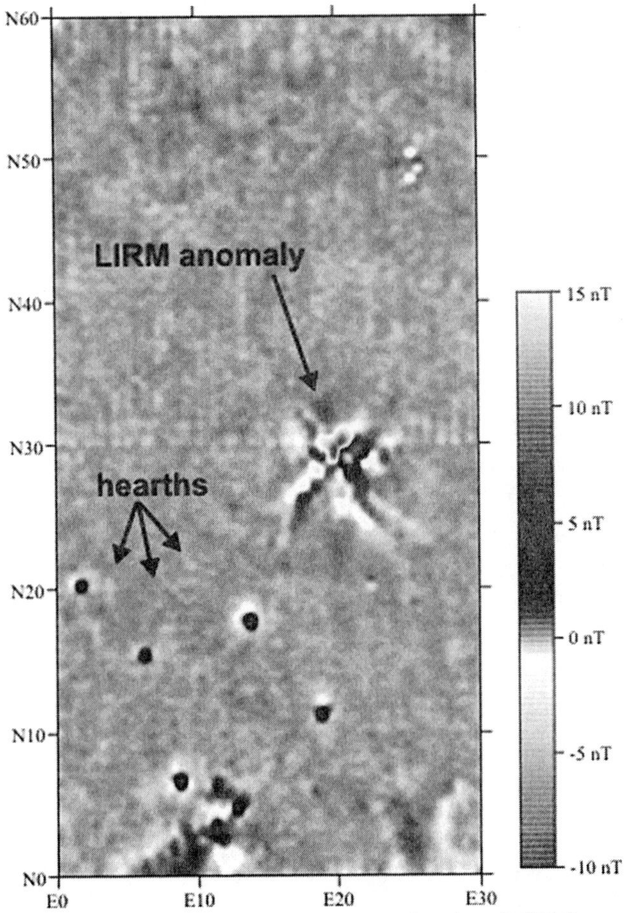

Figure 8: Lightning induced remanent magnetisation (LIRM) mapped during a magnetic field gradient survey of an archaeological site located in Wyoming, United States

The movement of electrical charges produces a magnetic field. The intense currents of a lightning discharge create a fleeting but very strong magnetic field. Where the lightning current path passes through rock, soil, or metal these materials can become permanently magnetised. This effect is known as lightning-induced remanent magnetism, or LIRM. These currents follow the least resistive path, often horizontally near the surface but sometimes vertically, where faults, ore bodies, or ground water offers a less resistive path. Lightning-induced magnetic anomalies can be mapped in the ground, and analysis of magnetised materials can confirm lightning was the source of the magnetisation and provide an estimate of the peak current of the lightning discharge.

Records and Locations

An old estimate of the frequency of lightning on Earth was 100 times a second. Now that there are satellites that can detect lightning, including in places where there is nobody to observe it, it is known to occur on average 44 ± 5 times a second, for a total of nearly 1.4 billion flashes per year; 75% of these flashes are either cloud-to-cloud or intra-cloud and 25% are cloud-to-ground.

The maps on the right show that lightning is not distributed evenly around the planet. Approximately 70% of lightning occurs in the tropics where the majority of thunderstorms occur. The place where lightning occurs most often (according to the data from 2004–2005) is near the small village of Kifuka in the mountains of eastern Democratic Republic of the Congo, where the elevation is around 975 metres (3,200 ft). On average this region receives 158 lightning strikes per square kilometre (approx. 0.4 square mile) a year. Above the Catatumbo river, which feeds Lake Maracaibo in Venezuela Catatumbo lightning flashes several times per minute, 140 to 160 nights per year, accounting for 25% of the world's production of upper-atmospheric ozone. Singapore has one of the highest rates of lightning activity in the world. The city of Teresina in northern Brazil has the third-highest rate of occurrences of lightning strikes in the world. The surrounding region is referred to as the *Chapada do Corisco* ("Flash Lightning Flatlands"). In the US, Central Florida sees more lightning than any other area. For example, in what is called "Lightning Alley", an area from Tampa, to Orlando, there are as many as 50 strikes per square mile (about 20 per km^2) per year. The Empire State Building is struck by lightning on average 23 times each year, and was once struck 8 times in 24 minutes.

- Roy Sullivan held a Guinness World Record after surviving 7 different lightning strikes across 35 years.
- In July 2007, lightning killed up to 30 people when it struck a remote mountain village Ushari Dara in northwestern Pakistan.

- On 31 October 2005, sixty-eight dairy cows, all in full milk, died on a farm at Fernbrook on the Waterfall Way near Dorrigo, New South Wales after being struck by lightning. Three others were paralysed for several hours but they later made a full recovery. The cows were sheltering under a tree when it was struck by lightning and the electricity spread onto the surrounding soil killing the animals.

Lightning rarely strikes the open ocean, although some sea regions are lightning "hot spots". Winter storms passing off the east coast of the United States often erupt with electrical activity when they cross the warm waters of the Gulf Stream. The Gulf Stream endures about the same number of lightning strikes as the southern plains of the USA.

Lightning Detection

The earliest detector invented to warn of the approach of a thunder storm was the lightning bell. Benjamin Franklin installed one such device in his house. The detector was based on an electrostatic device called the 'electric chimes' invented by Andrew Gordon in 1742.

Lightning discharges generate a wide range of electromagnetic radiations, including radio-frequency pulses. The times at which a pulse from a given lightning discharge arrive at several receivers can be used to locate the source of the discharge. The United States federal government has constructed a nation-wide grid of such lightning detectors, allowing lightning discharges to be tracked in real time throughout the continental U.S.

In addition to ground-based lightning detection, several instruments aboard satellites have been constructed to observe lightning distribution. These include the Optical Transient Detector (OTD), aboard the Orb View-1 satellite launched on April 3, 1995, and the subsequent Lightning Imaging Sensor (LIS) aboard TRMM launched on November 28, 1997.

Notable Lightning Strikes

Some lightning strikes have caused either numerous fatalities or great damage. The following is a partial list:

- In 1660, lightning struck the gunpowder magazine at Osaka Castle, Japan; the resultant explosion set the castle on fire. In 1665, lightning struck the main tower of the castle and it burned down to the foundation.

- A particularly deadly lightning incident occurred in Brescia, Italy in 1769. Lightning struck the Church of St. Nazaire, igniting the 100 tons of gunpowder in its vaults; the resulting explosion killed 3000 people and destroyed a sixth of the city.
- 1902: A lightning strike damaged the upper section of the Eiffel Tower, requiring the reconstruction of its top
- December 8, 1963: Pan Am Flight 214 crashed as result of a lightning strike, killing all 81 people on board.
- July 12, 1970, the central mast of the Orlunda radio transmitter collapsed after a lightning strike destroyed its basement insulator.
- December 24, 1971: LANSA Flight 508 crashed as a result of lightning in Peru, with 91 people killed.
- November 2, 1994, lightning struck fuel tanks in Dronka, Egypt and caused 469 fatalities.

Harvesting Lightning Energy

Since the late 1980s there have been several attempts to investigate the possibility of harvesting energy from lightning. While a single bolt of lightning carries a relatively large amount of energy (approximately 5 billion joules), this energy is concentrated in a small location and is passed during an extremely short period of time (milliseconds); therefore, extremely high electrical power is involved. It has been proposed that the energy contained in lightning be used to generate hydrogen from water, or to harness the energy from rapid heating of water due to lightning.

A technology capable of harvesting lightning energy would need to be able to rapidly capture the high power involved in a lightning bolt. Several schemes have been proposed, but the ever-changing energy involved in each lightning bolt render lightning power harvesting from ground based rods impractical - too high, it will damage the storage, too low and it may not work. According to Northeastern University physicists Stephen Reucroft and John Swain, a lightning bolt carries a few million joules of energy, enough to power a 100-watt bulb for 5.5 hours. Additionally, lightning is sporadic, and therefore energy would have to be collected and stored; it is difficult to convert high-voltage electrical power to the lower-voltage power that can be stored.

In the summer of 2007, an alternative energy company called Alternate Energy Holdings, Inc. (AEHI) tested a method for capturing the energy in lightning bolts. The design for the system had been purchased from an Illinois inventor named Steve LeRoy, who had reportedly been able to power a 60-watt light bulb for 20 minutes using the energy captured from a small flash of artificial lightning. The method involved a tower, a means of shunting off a large portion of the incoming energy, and a capacitor to store the rest. According to Donald Gillispie, CEO of AEHI, they "couldn't make it work," although "given enough time and money, you could probably scale this thing up... it's not black magic; it's truly math and science, and it could happen."

According to Dr. Martin A. Uman, co-director of the Lightning Research Laboratory at the University of Florida and a leading authority on lightning, a single lightning strike, while fast and bright, contains very little energy, and dozens of lighting towers like those used in the system tested by AEHI would be needed to operate five 100-watt light bulbs for the course of a year. When interviewed by *The New York Times*, he stated that the energy in a thunderstorm is comparable to that of an atomic bomb, but trying to harvest the energy of lightning from the ground is "hopeless".

Another major challenge when attempting to harvest energy from lighting is the impossibility of predicting when and where thunderstorms will occur. Even during a storm, it is very difficult to tell where exactly lightning will strike.

A relatively easy method is the direct harvesting of atmospheric charge before it turns into lightning. At a small scale, it was done a few times with the most known example being Benjamin Franklin's experiment with his kite. However, to collect reasonable amounts of energy very large constructions are required, and it is relatively hard to utilise the resulting extremely high voltage with reasonable efficiency.

In Culture

As Expressions and Symbols: The expression "Lightning never strikes twice (in the same place)" is similar to "Opportunity never knocks twice" in the vein of a "once in a lifetime" opportunity, *i.e.*, something that is generally considered improbable. Lightning occurs frequently and more so in specific areas. Since various factors alter the probability of strikes at any given location, repeat lightning strikes have a very low probability (but are not impossible). Similarly, "A bolt from the blue" refers to something totally unexpected. Some political parties use lightning flashes as a symbol of power, such as the People's Action Partyin Singapore and the British Union of Fascists during the 1930s. The Schutzstaffel, the secret police of the Nazi Party, used the Sig rune in their logo which symbolises lightning. The German word Blitzkrieg, which means "lightning war", was a major offensive strategy of the German army during World War II.

In French and Italian, the expression for "Love at first sight" is *Coup de foudre* and *Colpo di fulmine*, respectively, which literally translated means "lightning strike". Some European languages have a separate word for lightning which strikes the ground (as opposed to lightning in general); often it is a cognate of the English word "rays". The name of New Zealand's most celebrated thoroughbred horse,Phar Lap, derives from the shared Zhuang and Thai word for lightning.

The bolt of lightning in heraldry is called a thunderbolt and is shown as a zigzag with non-pointed ends. This symbol usually represents power and speed. In Hindu mythology the thunderbolt (Sanskrit *Vajra*) is an attribute of the Hindu god Indra. The lightning bolt or thunderbolt appears also as a heraldic charge. The lightning bolt is used to represent the instantaneous communication capabilities of electrically-powered telegraphs and radios, and is a common insignia for military communications units throughout the world. A lightning bolt is also the NATO symbol for a signal asset.

Ceraunoscopy

This is divination by observing lightning or by listening to thunder. It is a type of aeromancy. People have also sought to control lightning by conducting rituals or casting spells.

Religion

Over the centuries, lightning in cultures was viewed as part of a deity or a deity in of itself. One of the most classic portrayals of this is of the Greek god Zeus. An ancient story is when Zeus was at war against Kronus and the Titans, he released his brothers, Hades and Poseidon, along with the Cyclopes. In turn, the Cyclopes gave Zeus the thunderbolt as a weapon, which was near the beginning of Zeus himself. The thunderbolt became a popular symbol of Zeus and continues to be today.

The Aztecs portrayed lightning as a supernatural power of the god Tlaloc, visualised as his axe. In mythology, Tlaloc was the bringer not only of beneficial rain but of storms, killer lightning bolts, flood, and disease. The Classic Mayas personified lightning as a rain deity classified by scholars as God K. This deity has a leg shaped like a lightning serpent, and a forehead perforated by a lightning celt. A miniature God K is often wielded as an axe by the king.

In Slavic mythology the highest god of the pantheon is Perun, the god of thunder and lightning.

Pçrkons/Perkûnas is the common Baltic god of thunder, one of the most important deities in the Baltic pantheon. In both Latvian and Lithuanian mythology, he is documented as the god of thunder, rain, mountains, oak trees and the sky. In Norse mythology, Thor is the god of thunder and the sound of thunder comes from the chariot he rides across the sky. The lightning comes from his hammer Mjölnir.

In Finnish mythology, Ukko (engl. *Old Man*) is the god of thunder, sky and weather. The Finnish word for thunder is *ukkonen*, derived from the god's name.

In the Jewish religion, a blessing *"...He who does acts of creation"* is to be recited, upon sighting lightning. The Talmud refers to the Hebrew word for the sky, ("Shamaim") - as built from fire and water ("Esh Umaim"), since the sky is the source of the inexplicable mixture of "fire" and water that come together, during rainstorms. This is mentioned in various prayers and discussed in writings of Kabbalah.

In Islam, the Quran states: *"He it is Who showeth you the lightning, a fear and a hope, and raiseth the heavy clouds. The thunder hymneth His praise and (so do) the angels for awe of Him. He launcheth the thunder-bolts and smiteth with them whom He will."* (Qur'an 13:12–13) and, *"Have you not seen how God makes the clouds move gently, then joins them together, then makes them into a stack, and then you see the rain come out of it..."* (Quran, 24:43). The preceding verse, after mentioning clouds and rain, speaks about hail and lightning, *"...And He sends down hail from mountains (clouds) in the sky, and He strikes with it whomever He wills, and turns it from whomever He wills."*

In India, the Hindu god Indra is considered the god of rains and lightning and the king of the Devas.

In Japan, the Shinto god Raijin is considered the god of lightning and thunder. He is depicted as a demon who strikes a drum to create lightning.

In the traditional religion of the African Bantu tribes, such as the Baganda and Banyoro of Uganda, lightning is a sign of the ire of the gods. The Baganda specifically attribute the lightning phenomenon to the god Kiwanuka, one of the main trio in the Lubaale gods of the sea or lake. Kiwanuka starts wild fires, strikes trees and other high buildings, and a number of shrines are established in the hills, mountains and plains to stay in his favor. Lightning is also known to be invoked upon one's enemies by uttering certain chants, prayers, and making sacrifices.

Ball Lightning

Ball lightning is an unexplained atmospheric electrical phenomenon. The term refers to reports of luminous, usually spherical objects which vary from pea-sized to several metres in diameter. It is usually associated with thunderstorms, but lasts considerably longer than the split-second flash of a lightning bolt. Many of the early reports say that the ball eventually explodes, sometimes with fatal consequences, leaving behind the odour of sulphur.

Laboratory experiments have produced effects that are visually similar to reports of ball lightning, but it is presently unknown whether these are actually related to any naturally occurring phenomenon. Scientific data on natural ball lightning are scarce owing to its infrequency and unpredictability. The presumption of its existence is based on reported public sightings, and has therefore produced somewhat inconsistent findings. Given inconsistencies and the lack of reliable data, the true nature of ball lightning is still unknown.

Historical Accounts

In a 1960 study, 5% of the US population reported having witnessed ball lightning. Another study analysed reports of 10,000 cases.

Figure 9: A contemporary woodcut of the 1638 thunderstorm at Widecombe

M. l'abbé de Tressan, in *Mythology compared with history: or, the fables of the ancients elucidated from historical records*:

... during a storm which endangered the ship Argo, fires were seen to play round the heads of the Tyndarides, and the instant after the storm ceased. From that time, those fires which frequently appear on the surface of the ocean were called the fire of Castor and Pollux. When two were seen at the same time, it announced the return of calm, when only one, it was the presage of a dreadful storm. This species of fire is frequently seen by sailors, and is a species of *ignis fatuus*.

The Great Thunderstorm of Widecombe-in-the-Moor

One of the earliest descriptions was reported during The Great Thunderstorm at a church in Widecombe-in-the-Moor, Devon, in England, on 21 October 1638. Four people died and approximately 60 were injured when, during a severe storm, an 8-foot (2.4 m) ball of fire was described as striking and entering the church, nearly destroying it. Large stones from the church walls were hurled into the ground and through large wooden beams. The ball of fire allegedly smashed the pews and many windows, and filled the church with a foul sulphurous odour and dark, thick smoke.

The ball of fire reportedly divided into two segments, one exiting through a window by smashing it open, the other disappearing somewhere inside the church. The explanation at the time, because of the fire and sulphur smell, was that the ball of fire was "the devil" or the "flames of hell". Later, some blamed the entire incident on two people who had been playing cards in the pew during the sermon, thereby incurring God's wrath.

The Catherine and Mary

In December 1726 a number of British newspapers printed an extract of a letter from John Howell of the sloop *Catherine and Mary*:

As we were coming thro' the Gulf of Florida on the 29th of August, a large ball of fire fell from the Element and split our mast in Ten Thousand Pieces, if it were possible; split our Main Beam, also Three Planks of the Side, Under Water, and Three of the Deck; killed one man, another had his Hand carried of, [sic] and had it not been for the violent rains, our Sails would have been of a Blast of Fire.

The Montague

One particularly large example was reported "on the authority of Dr. Gregory" in 1749:

Admiral Chambers on board the Montague, November 4, 1749, was taking an observation just before noon...he observed a large ball of blue fire about three miles distant from them. They immediately lowered their topsails, but it came up so fast upon them, that, before they could raise the main tack, they observed the ball rise almost perpendicularly, and not above forty or fifty yards from the main chains when it went off with an explosion, as great as if a hundred cannons had been discharged at the same time, leaving behind it a strong sulphurous smell. By this explosion the main top-mast was shattered into pieces and the main mast went down to the keel. Five men were knocked down and one of them much bruised. Just before the explosion, the ball seemed to be the size of a large mill-stone.

Georg Richmann

A 1753 report depicts ball lightning as being lethal, when Professor Georg Richmann of Saint Petersburg, Russia, created a kite-flying apparatus similar to Benjamin Franklin's proposal a year earlier. Richmann was attending a meeting of the Academy of Sciences when he heard thunder, and ran home with his engraver to capture the event for posterity. While the experiment was under way, ball lightning appeared and travelled down the string, struck Richmann's forehead and killed him. The ball left a red spot on Richmann's forehead, his shoes were blown open, and his clothing was singed. His engraver was knocked unconscious. The door frame of the room was split and the door was torn from its hinges.

HMS Warren Hastings

An English journal reported that during an 1809 storm, three "balls of fire" appeared and "attacked" the British ship HMS *Warren Hastings*. The crew watched one ball descend, killing a man on deck and setting the main mast on fire. A crewman went out to retrieve the fallen body and was struck by a second ball, which knocked him back and left him with mild burns. A third man was killed by contact with the third ball. Crew members reported a persistent, sickening sulphur smell afterward.

Ebenezer Cobham Brewer

Ebenezer Cobham Brewer, in his 1864 US edition of *A Guide to the Scientific Knowledge of Things Familiar*, discussed "globular lightning". He describes it as slow-moving balls of fire or explosive gas that sometimes fall to the earth or run along the ground during a thunderstorm. He said that the balls sometimes split into smaller balls and may explode "like a cannon".

Wilfrid de Fonvielle

In his book *Thunder and Lighting*, translated into English in 1875, French science writer, Wilfred de Fonvielle, wrote that there had been about 150 reports of globular lightning:

Globular lighting seems to be particularly attracted to metals; thus it will seek the railings of balconies, or else water or gas pipes etc, It has no peculiar tint of its own but will appear of any colour as the case may be...at Coethen in the Duchy of Anhalt it appeared green. M. Colon, Vice-President of the Geological Society of Paris, saw a ball of lightning descend slowly from the sky along the bark of a poplar tree; as soon as it touched the earth it bounced up again, and disappeared without exploding. On 10th of September 1845 a ball of lightning entered the kitchen of a house in the village of Salagnac in the valley of Correze. This ball rolled across without doing any harm to two women and a young man who were here; but on getting into an adjoining stable it exploded and killed a pig which happened to be shut up there, and which, knowing nothing about the wonders of thunder and lightning, dared to smell it in the most rude and unbecoming manner. The motion of such balls is far from being very rapid — they have even been observed occasionally to pause in their course, but they are not the less destructive for all that. A ball of lightning which entered the church of Stralsund, on exploding, projected a number of balls which exploded in their turn like shells.

Tsar Nicholas II

Tsar Nicholas II, the last Emperor of Russia, reported witnessing what he called "a fiery ball" while in the company of his grandfather, Tsar Alexander II: "Once my parents were away," recounted the Tsar, "and I was at the all-night vigil with my grandfather in the small church in Alexandria. During the service there was a powerful thunderstorm, streaks of lightning flashed one after the other, and it seemed as if the peals of thunder would shake even the church and the whole world to its foundations. Suddenly it became quite dark, a blast of wind from the open door blew out the flame of the candles which were lit in front of the iconostasis, there was a long clap of thunder, louder than before, and I suddenly saw a fiery ball flying from the window straight towards the head of the Emperor. The ball (it was of lightning) whirled around the floor, then passed the chandelier and flew out through the door into the park. My heart froze, I glanced at my grandfather – his face was completely calm. He crossed himself just as calmly as he had when the fiery ball had flown near us, and

I felt that it was unseemly and not courageous to be frightened as I was ... After the ball had passed through the whole church, and suddenly gone out through the door, I again looked at my grandfather. A faint smile was on his face, and he nodded his head at me. My panic disappeared, and from that time I had no more fear of storms."

Aleister Crowley

British occultist Aleister Crowley reported witnessing what he referred to as "globular electricity" during a thunderstorm on Lake Pasquaney in New Hampshire in 1916. He was sheltered in a small cottage when he "noticed, with what I can only describe as calm amazement, that a dazzling globe of electric fire, apparently between six and twelve inches (15–30 cm) in diameter, was stationary about six inches below and to the right of my right knee. As I looked at it, it exploded with a sharp report quite impossible to confuse with the continuous turmoil of the lightning, thunder and hail, or that of the lashed water and smashed wood which was creating a pandemonium outside the cottage. I felt a very slight shock in the middle of my right hand, which was closer to the globe than any other part of my body."

Other Accounts

- On 30 April 1877, a ball of lightning entered the Golden Temple at Amritsar, India, and exited through a side door. Several people observed the ball, and the incident is inscribed on the front wall of Darshani Deodhi.
- On 22 November 1894 there was an unusually prolonged instance of natural ball lightning in Golden, Colorado which suggests it could be artificially induced from the atmosphere. The *Golden Globe* newspaper reported "A beautiful yet strange phenomenon was seen in this city on last Monday night. The wind was high and the air seemed to be full of electricity. In front of, above and around the new Hall of Engineering of the School of Mines, balls of fire played tag for half an hour, to the wonder and amazement of all who saw the display. In this building is situated the dynamos and electrical apparatus of perhaps the finest electrical plant of its size in the state. There was probably a visiting delegation from the clouds, to the captives of the dynamos on last Monday night, and they certainly had a fine visit and a roystering game of romp."
- In July 1907 the Cape Naturaliste Lighthouse in Western Australia was hit by ball lightning. Lighthouse keeper

Patrick Baird was in the tower at the time and was knocked unconscious. His daughter Ethel recorded the event.

- An early fictional reference to ball lightning appears in a children's book set in the 19th century by Laura Ingalls Wilder. The books are considered historical fiction, but the author always insisted they were descriptive of actual events in her life. In Wilder's description, three separate balls of lightning appear during a winter blizzard near a cast iron stove in the family's kitchen. They are described as appearing near the stovepipe, then rolling across the floor, only to disappear as the mother (Caroline Ingalls) chases them with a willow-branch broom.
- Pilots in World War II described an unusual phenomenon for which ball lightning has been suggested as an explanation. The pilots saw small balls of light moving in strange trajectories, which came to be referred to as foot fighters.
- Submariners in WWII gave the most frequent and consistent accounts of small ball lightning in the confined submarine atmosphere. There are repeated accounts of inadvertent production of floating explosive balls when the battery banks were switched in or out, especially if mis-switched or when the highly inductive electrical motors were mis-connected or disconnected. An attempt later to duplicate those balls with a surplus submarine battery resulted in several failures and an explosion.
- On 6 August 1944, a ball of lightning went through a closed window in Uppsala, Sweden, leaving a circular hole about 5 cm in diameter. The incident was witnessed by residents in the area, and was recorded by a lightning strike tracking system on the Division for Electricity and Lightning Research at Uppsala University.
- In 1954 Domokos Tar, a physicist, observed a lightning strike during a heavy thunderstorm. A single bush was flattened in the wind. Some seconds later a speedy rotating ring (cylinder) appeared in the shape of a wreath. The ring was about 5 m away from the lightning impact point. The ring's plane was perpendicular to the ground and in full view of the observer. The outer/inner diameters were about 60/30 cm. The ring rotated quickly about 80 cm above the ground. It was composed

of wet leaves and dirt and rotated counter clockwise. After seconds the ring became self-illuminated turning increasingly red, then orange, yellow and finally white. The ring (cylinder) at the outside was similar to a sparkler. In spite of the rain, many electrical high voltage discharges could be seen. After some seconds , the ring suddenly disappeared and simultaneously the Ball Lightning appeared in the middle. Initially the ball had only one tail and it rotated in the same direction as the ring. It was homogenous and showed no transparency. In the first moment the ball hovered motionless, but then began to move forward on the same line with a constant speed of about 1m/sec. It was stable and travelled at the same height in spite of the heavy rain and strong wind. After moving about 10 m it suddenly disappeared without any noise.

- On 10 July 2011, during a powerful thunderstorm, a ball of light with a two-metre tail went through a window to the control room of local emergency services in Liberec, Czech Republic. The ball bounced from window to the ceiling, then to the floor and back to the ceiling, where it rolled along it for two or three metres. Then it dropped to the floor and disappeared. The staff present in the control room was frightened, smelled electricity and burned cables and thought something was burning. The computers froze (not crashed) and all communications equipment was knocked out for the night until restored by technicians. Aside from damages caused by disrupting equipment, only one computer monitor was destroyed.

Characteristics

Descriptions of ball lightning vary wildly. It has been described as moving up and down, sideways or in unpredictable trajectories, hovering and moving with or against the wind; attracted to, unaffected by, or repelled from buildings, people, cars and other objects. Some accounts describe it as moving through solid masses of wood or metal without effect, while others describe it as destructive and melting or burning those substances. Its appearance has also been linked to power lines as well as during thunderstorms and also calm weather. Ball lightning has been described as transparent, translucent, multicoloured, evenly lit, radiating flames, filaments or sparks, with shapes that vary between spheres, ovals, tear-drops, rods, or disks.

Ball lightning is often erroneously identified as St. Elmo's fire. They are separate and distinct phenomena.

The balls have been reported to disperse in many different ways, such as suddenly vanishing, gradually dissipating, absorption into an object, "popping," exploding loudly, or even exploding with force, which is sometimes reported as damaging. Accounts also vary on their alleged danger to humans, from lethal to harmless.

A review of the available literature published in 1972 identified the properties of a "typical" ball lightning, whilst cautioning against over-reliance on eye-witness accounts:

- They frequently appear almost simultaneously with cloud-to-ground lightning discharge
- They are generally spherical or pear-shaped with fuzzy edges
- Their diameters range from 1–100 cm, most commonly 10–20 cm
- Their brightness corresponds to roughly that of a domestic lamp, so they can be seen clearly in daylight
- A wide range of colours has been observed, red, orange and yellow being the most common.
- The lifetime of each event is from 1 second to over a minute with the brightness remaining fairly constant during that time
- They tend to move, most often in a horizontal direction at a few metres per second, but may also move vertically, remain stationary or wander erratically.
- Many are described as having rotational motion
- It is rare that observers report the sensation of heat, although in some cases the disappearance of the ball is accompanied by the liberation of heat
- Some display an affinity for metal objects and may move along conductors such as wires or metal fences
- Some appear within buildings passing through closed doors and windows
- Some have appeared within metal aircraft and have entered and left without causing damage
- The disappearance of a ball is generally rapid and may be either silent or explosive

- Odours resembling ozone, burning sulphur, or nitrogen oxides are often reported

Laboratory Experiments

Scientists have long attempted to produce ball lightning in laboratory experiments. While some experiments have produced effects that are visually similar to reports of natural ball lightning, it has not yet been determined whether there is any relation.

Nikola Tesla was reportedly able to artificially produce 1.5" (3.8 cm) balls and conducted some demonstrations of his ability, but he was really interested in higher voltages and powers, and remote transmission of power, so the balls he made were just a curiosity.

The International Committee on Ball Lightning holds regular symposia on the subject, the most recent of which took place in Kaliningrad, Russia in 2008. A related group uses the generic name "Unconventional Plasmas".

Water Discharge Experiments

Some scientific groups, including the Max Planck Institute, have reportedly produced a ball lightning-type effect by discharging a high-voltage capacitor in a tank of water.

Home Microwave Oven Experiments

Many modern experiments involve using a microwave oven to produce small rising glowing balls, often referred to as "plasma balls". Generally, the experiments are conducted by placing a lit or recently extinguished match or other small object in a microwave oven. The burnt portion of the object flares up into a large ball of fire, while "plasma balls" can be seen floating near the ceiling of the oven chamber. Some experiments describe covering the match with an inverted glass jar, which contains both the flame and the balls so that they will not damage the chamber walls. Experiments by Eli Jerby and Vladimir Dikhtyar in Israel revealed that microwave plasma balls are made up of nanoparticles with an average radius of 25 nm. The Israeli team demonstrated the phenomenon with copper, salts, water and carbon.

Silicon Experiments

Experiments in 2007 involved shocking silicon wafers with electricity, which vaporises the silicon and induces oxidation in the vapours. The visual effect can be described as small glowing, sparkling orbs that roll around a surface. Two Brazilian scientists, Antonio Pavão and Gerson Paiva of the Federal University of Pernambuco have reportedly consistently made small long-lasting balls using this method. These experiments stemmed from the theory that ball lightning is actually oxidised silicon vapours.

Transcranial Magnetic Stimulation Analogy

Theoretical calculations from University of Innsbruck researchers suggest that the magnetic fields involved in certain types of lightning strikes could potentially induce visual hallucinations resembling ball lightning. Such fields, which are found within close distances to a point in which multiple lightning strikes have occurred over a few seconds, can directly cause the neurons in the visual cortexto fire, resulting in magnetophosphenes (magnetically-induced visual hallucinations).

Possible Scientific Explanations

An attempt to explain ball lightning was made by Nikola Tesla in 1904, but there is at present no widely-accepted explanation for the phenomenon. Several theories have been advanced since it was brought into the scientific realm by the English physician and electrical researcher William Snow Harris in 1843, and French Academy scientist François Arago in 1855.

Microwave Cavity Hypothesis

Pyotr Kapitsa proposed that ball lightning is a glow discharge driven by microwave radiation that is guided to the ball along lines of ionised air from lightning clouds where it is produced. The ball serves as a resonant microwave cavity, automatically adjusting its radius to the wavelength of the microwave radiation so that resonance is maintained.

The Handel Maser-Soliton theory of ball lightning hypothesizes that the energy source generating the ball lightning is a large (several cubic kilometres) atmospheric maser. The ball lightning appears as a plasma caviton at the antinodal plane of the microwave radiation from the maser.

Soliton Hypothesis

Julio Rubenstein, David Finkelstein, and James R. Powell proposed that ball lightning is a detached St. Elmo's fire (1964-1970). St. Elmo's fire arises when a sharp conductor, such as a ship's mast, amplifies the atmospheric electric field to breakdown. For a globe the amplification factor is 3. A free ball of ionised air can amplify the ambient field this much by its own conductivity. When this maintains the ionisation, the ball is then a soliton in the flow of atmospheric electricity. Powell's kinetic theory calculation found that the ball size is set by the second Town send coefficient (the mean free path of conduction electrons) near breakdown. Wandering glow discharges are found to occur within certain industrial microwave ovens and continue to glow for several seconds after power is shut off. Arcs drawn from high-power low-voltage microwave generators also are found to exhibit after-glow. Powell measured their spectra and found the after-glow to come mostly from metastable NO ions, which are long-lived at low temperatures. It occurred in air and in nitrous oxide, which possess such metastable ions, and not in atmospheres of argon, carbon dioxide, or helium, which do not.

Vaporised Silicon Hypothesis

This hypothesis suggests that ball lightning consists of vaporised silicon burning through oxidation. Lightning striking Earth's soil could vaporise the silica contained within it, turning it into pure silicon vapour. As it cools, the silicon could condense into a floating aerosol, bound by its charge, glowing due to the heat of silicon recombining with oxygen. An experimental investigation of this effect, published in 2007, reported producing "luminous balls with lifetime in the order of seconds" by evaporating pure silicon with an electric arc. Videos of this experiment have been made available.

Aerodynamic Vortex is Cut Causing it to Shrink into a Sphere Hypothesis

According to his ball lightning observation, physicist Domokos Tar suggests the following theory for ball lightning formation. Lightning strikes perpendicular to the ground. The thunder, which contains more than 99.9% of the lightning energy, follows immediately at supersonic speed in the form of shock waves and forms an invisible aerodynamic turbulence ring lying horizontal to the ground. Around the ring there is an over and under pressure which rotates the vortex around its circular axis in the cross section of the torus. At the same time, the ring expands concentrically parallel to the ground at low speed. In an open space the vortex fades and finally disappears. If the vortex's expansion is obstructed and symmetry is broken, the vortex breaks into a sickle form, still invisible, and because of the central and surface tension-forces it shrinks through an intermediate state of a cylinder and finally into a ball. The whole energy of the turning vortex concentrates first in a turning linear-cylinder slowly becoming visible. The ball lightning has the same turning axis as the rotating cylinder. It is believed that the energy of the vortex ring is about million times less than the energy of the thunder. The vortex, during shrinking, gives its full energy to the ball. In some observations the ball has had an extremely high energy but this phenomenon is not yet clear.

The present theory concerns only the low energy lightning ball form, where there must be a spherical form with centripetal forces and surface tension. Practically the whole energy of the vortex is concentrated in the ball according to the law of conservation of mass, momentum and energy. The illumination of the cylinder and later of the ball is caused by triboelectricity and electrolumine scence. Many sparklers on the outside of the cylinder seen during the observation, proved this. The sparkler's direction indicated the cylinder's turning direction. This proves that the ball was not created from the lightning's channel material because according to the law of laminar flow, if the ball came from the channel it would have turned in the opposite direction. Therefore, the BL is created from dirt, leaves and other particles in the air. The LB has H2O, CO2, O2, N2, sulphur etc excited radiating molecules.

Nanobattery Hypothesis

Oleg Meshcheryakov suggests that ball lightning is made of composite nano or submicrometre particles, each particle constituting a battery. A surface discharge shorts these batteries, resulting in a current which forms the ball. His model is described as an aerosol, but not aerogel, model that explains all the observable properties and processes of ball lightning.

Black Hole Hypothesis

Another hypothesis is that some ball lightning is the passage of microscopic primordial black holes through the Earth's atmosphere as proposed by Mario Rabinowitz in Astrophysics and Space Science journal in 1999. Inspired by M. Fitzgerald's account of ball lightning on 6 August 1868, in Ireland that lasted 20 minutes and left a 6 metre square hole, a 90 metre long trench, a second trench 25 meters long, and a small cave in the peat bog, Pace Van Devender, a plasma physicist at Sandia National Laboratories in Albuquerque, New Mexico, and his team found depressions consistent with Fitzgerald's report and inferred that the evidence is inconsistent with thermal (chemical or nuclear) and electrostatic effects. An electromagnetically levitated, compact mass of over 20,000 kg would produce the reported effects but requires a density of more than 2000 times the density of gold, which implies a miniature black hole. He and his team found a second event in the peat-bog witness plate from 1982 and are currently trying to geolocate electromagnetic emission consistent with the hypothesis. His colleagues at the institute agreed that, implausible though the hypothesis seemed, it was worthy of their attention.

Buoyant Plasma Hypothesis

The declassified Project Condign report concludes that buoyant charged plasma formations similar to ball lightning are formed by novel physical, electrical, and magnetic phenomena, and that these charged plasmas are capable of being transported at enormous speeds under the influence and balance of electrical charges in the atmosphere. These plasmas appear to originate due to more than one set of weather and electrically-charged conditions, the scientific rationale for which is incomplete or not fully understood. One suggestion is that meteors breaking up in the atmosphere and forming charged plasmas as opposed to burning completely or impacting as meteorites could explain some instances of the phenomena, in addition to other unknown atmospheric events.

Transcranial Magnetic Stimulation

Cooray and Cooray (2008) stated that the features of hallucinations experienced by patients having epileptic seizures in the occipital lobe are similar to the observed features of ball lightning. The study also showed that the rapidly changing magnetic field of a close lightning flash has a strength which is large enough to excite the neurons in the brain strengthening the possibility of lightning-induced seizure in the occipital lobe of a person located close to a lightning strike establishing the connection between epileptic hallucination mimicking ball lightning and thunderstorms. More recent research with transcranial magnetic stimulation has been shown to give the same hallucination results in the laboratory (termed magnetophospenes), and these conditions have been shown to occur in nature near lightning strikes.

Other Hypotheses

Several other hypotheses have been proposed to explain ball lightning:

- Spinning electric dipole hypothesis. A 1976 article by V. G. Endean postulated that ball lightning could be described as an electric field vector spinning in the microwave frequency region.

- Electrostatic Leyden jar models. Stanley Singer discussed (1971) this type of hypothesis and suggested that the electrical recombination time would be too short for the ball lightning lifetimes often reported.
- J. Pace Van Devender separates extreme ball lightning of the highly energetic violent kind, and proposes a theory of neutrinos and heavy neutrinos.
- Smirnov proposed (1987) a fractal aerogel hypothesis.
- V.P. Torchigin proposed (2003) considering ball lightning as a form of self-confined intense light.
- M.I. Zelikin proposed (2006) an explanation (with strict mathematical background) based on the hypothesis of plasma superconductivity.
- Ph. M. Papaelias studied (1984) the antimatter meteor hypothesis as a possible explanation of ball lightning formation. He compared all properties of ball lightning to those expected by antimatter meteor undergoing annihilation by atmospheric molecules and found almost identical properties.

Ionosphere

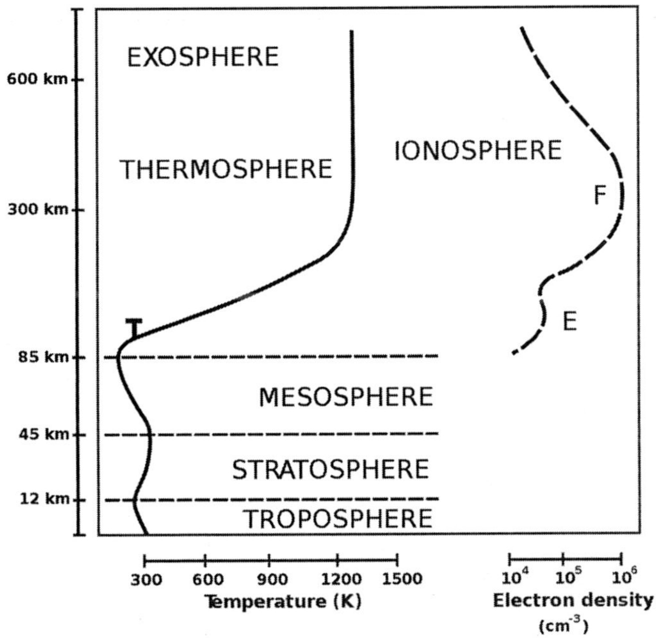

Figure 10: Relationship of the atmosphere and ionosphere

The ionosphere is a part of the upper atmosphere, comprising portions of the mesosphere, thermosphere and exosphere, distinguished because it is ionised by solar radiation. It plays an important part in atmospheric electricity and forms the inner edge of the magnetosphere. It has practical importance because, among other functions, it influences radio propagation to distant places on the Earth.

Geophysics

The ionosphere is a shell of electrons and electrically charged atoms and molecules that surrounds the Earth, stretching from a height of about 50 km to more than 1000 km. It owes its existence primarily to ultraviolet radiation from the Sun.

The lowest part of the Earth's atmosphere, thetroposphere extends from the surface to about 10 km (6.2 mi). Above 10 km is the stratosphere, followed by the mesosphere. In the stratosphere incoming solar radiation creates the ozone layer. At heights of above 80 km (50 mi), in the thermosphere, the atmosphere is so thin that free electrons can exist for short periods of time before they are captured by a nearby positive ion. The number of these free electrons is sufficient to affect radio propagation. This portion of the atmosphere is *ionised* and contains a plasma which is referred to as the ionosphere. In a plasma, the negative free electrons and the positive ions are attracted to each other by the electromagnetic force, but they are too energetic to stay fixed together in an electrically neutral molecule.

Ultraviolet (UV), X-Ray and shorter wavelengths of solar radiation are *ionising,* since photons at these frequencies contain sufficient energy to dislodge an electron from a neutral gas atom or molecule upon absorption. In this process the light electron obtains a high velocity so that the temperature of the created electronic gas is much higher (of the order of thousand K) than the one of ions and neutrals. The reverse process to Ionisation is recombination, in which a free electron is "captured" by a positive ion, occurs spontaneously. This causes the emission of a photon carrying away the energy produced upon recombination. As gas density increases at lower altitudes, the recombination process prevails, since the gas molecules and ions are closer together. The balance between these two processes determines the quantity of ionisation present.

Ionisation depends primarily on the Sun and its activity. The amount of ionisation in the ionosphere varies greatly with the amount of radiation received from the Sun. Thus there is a diurnal (time of day) effect and a seasonal effect. The local winter hemisphere is tipped away from the Sun, thus there is less received solar radiation. The activity of the Sun is associated with the sunspot cycle, with more radiation occurring with more sunspots. Radiation received also varies with geographical location (polar, auroral zones, mid-latitudes, and equatorial regions). There are also mechanisms that disturb the ionosphere and decrease the ionisation. There are disturbances such as solar flares and the associated release of charged particles into the solar wind which reaches the Earth and interacts with its geomagnetic field.

The Ionospheric Layers

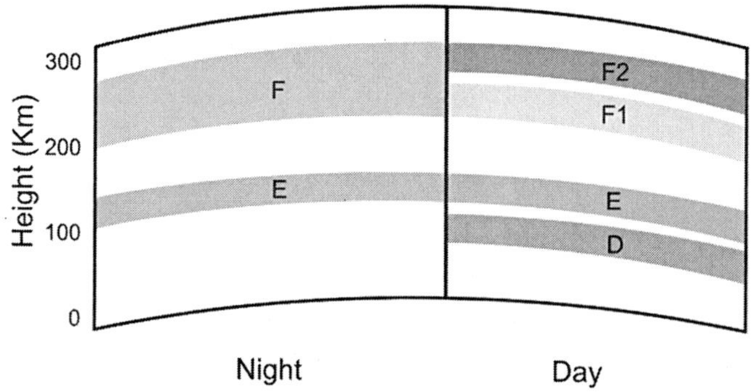

Figure 11: Ionospheric layers.

At night the F layer is the only layer of significant ionisation present, while the ionisation in the E and D layers is extremely low. During the day, the D and E layers become much more heavily ionised, as does the F layer, which develops an additional, weaker region of ionisation known as the F_1 layer. The F_2 layer persists by day and night and is the region mainly responsible for the refraction of radio waves.

D Layer

The D layer is the innermost layer, 60 km to 90 km above the surface of the Earth. Ionisation here is due to Lyman series-alpha hydrogen radiation at a wavelength of 121.5 nanometre (nm) ionising nitric oxide (NO). In addition, with high Solar activity hard X-rays (wavelength < 1 nm) may ionise (N_2, O_2). During the night cosmic rays produce a residual amount of ionisation. Recombination is high in the D layer, the net ionisation effect is low, but loss of wave energy is great due to frequent collisions of the electrons (about ten collisions every msec). As a result high-frequency (HF) radio waves are not reflected by the D layer but suffer loss of energy therein. This is the main reason for absorption of HF radio waves, particularly at 10 MHz and below, with progressively smaller absorption as the frequency gets higher. The absorption is small at night and greatest about midday. The layer reduces greatly after sunset; a small part remains due to galactic cosmic rays. A common example of the D layer in action is the disappearance of distant AM broadcast band stations in the daytime.

During solar proton events, ionisation can reach unusually high levels in the D-region over high and polar latitudes. Such very rare events are known as Polar Cap Absorption (or PCA) events, because the increased ionisation significantly enhances the absorption of radio signals passing through the region. In fact, absorption levels can increase by many tens of dB during intense events, which is enough to absorb most (if not all) transpolar HF radio signal transmissions. Such events typically last less than 24 to 48 hours.

E Layer

The E layer is the middle layer, 90 km to 120 km above the surface of the Earth. Ionisation is due to soft X-ray (1-10 nm) and far ultraviolet (UV) solar radiation ionisation of molecular oxygen (O_2). Normally, at oblique incidence, this layer can only reflect radio waves having frequencies lower than about 10 MHz and may contribute a bit to absorption on frequencies above. However, during intense Sporadic E events, the E_s layer can reflect frequencies up to 50 MHz and higher. The vertical structure of the E layer is primarily determined by the competing effects of ionisation and recombination. At night the E layer rapidly disappears because the primary source of ionisation is no longer present. After sunset an increase in the height of the E layer maximum increases the range to which radio waves can travel by reflection from the layer.

This region is also known as the Kennelly-Heaviside Layer or simply the Heaviside layer. Its existence was predicted in 1902 independently and almost simultaneously by the American electrical engineer Arthur Edwin Kennelly (1861–1939) and the British physicist Oliver Heaviside (1850–1925). However, it was not until 1924 that its existence was detected by Edward V. Appleton.

E_s

The E_s layer (sporadic E-layer) is characterised by small, thin clouds of intense ionisation, which can support reflection of radio waves, rarely up to 225 MHz. Sporadic-E events may last for just a few minutes to several hours. Sporadic E propagation makes radio amateurs very excited, as propagation paths that are generally unreachable can open up. There are multiple causes of sporadic-E that are still being pursued by researchers. This propagation occurs most frequently during the summer months when high signal levels may be reached. The skip distances are generally around 1,000 km (620 mi). Distances for one hop propagation can be as close as 900 km [500 miles] or up to 2,500 km (1,600 mi). Double-hop reception over 3,500 km (2,200 mi) is possible.

F Layer

The F layer or region, also known as the Appleton layer extends from about 200 km to more than 500 km above the surface of Earth. It is the densest point of the ionosphere, which implies signals penetrating this layer will escape into space. Beyond this layer is the topside ionosphere. Here extreme ultraviolet (UV, 10–100 nm) solar radiation ionises atomic oxygen. The F layer consists of one layer at night, but during the day, a deformation often forms in the profile that is labelled F_1. The F_2layer remains by day and night responsible for most skywave propagation of radio waves, facilitating high frequency (HF, or shortwave) radio communications over long distances.

From 1972 to 1975 NASA launched the AEROS and AEROS B satellites to study the F region.

Ionospheric Model

An ionospheric model is a mathematical description of the ionosphere as a function of location, altitude, day of year, phase of the sunspot cycle and geomagnetic activity. Geophysically, the state of the ionospheric plasma may be described by four parameters: *electron density, electron and iontemperature* and, since several species of ions are present, *ionic composition.* Radio propagationdepends uniquely on electron density.

Models are usually expressed as computer programs. The model may be based on basic physics of the interactions of the ions and electrons with the neutral atmosphere and sunlight, or it may be a statistical description based on a large number of observations or a combination of physics and observations. One of the most widely used models is the International Reference Ionosphere (IRI)(IRI 2007), which is based on data and specifies the four parameters just mentioned. The IRI is an international project sponsored by the Committee on

Space Research (COSPAR) and the International Union of Radio Science (URSI). The major data sources are the worldwide network of ionosondes, the powerful incoherent scatter radars (Jicamarca, Arecibo, Millstone Hill, Malvern, St. Santin), the ISIS and Alouette topside sounders, and in situ instruments on several satellites and rockets. IRI is updated yearly. IRI will be established in 2009 by the International Organisation for Standardisation (ISO) as standard TS16457. IRI is accurate in describing the variation of the electron density from bottom of the ionosphere to the altitude of maximum density than in describing the total electron content (TEC). Since 1999 the IRI is "International Standard" for the terrestrial ionosphere.

Anomalies to the Ideal Model

Ionograms allow deducing not only the shape of the different layers but also the structure of the electron/ion-plasma. Rough traces, indicating nonhomogeneity, are seen predominantly at night and at higher latitudes, and during disturbed conditions.

Winter Anomaly

At mid-latitudes, the F_2 layer daytime ion production is higher in the summer, as expected, since the Sun shines more directly on the Earth. However, there are seasonal changes in the molecular-to-atomic ratio of the neutral atmosphere that cause the summer ion loss rate to be even higher. The result is that the increase in the summertime loss overwhelms the increase in summertime production, and total F_2 ionisation is actually lower in the local summer months. This effect is known as the winter anomaly. The anomaly is always present in the northern hemisphere, but is usually absent in the southern hemisphere during periods of low solar activity.

Equatorial Anomaly

Figure 12: Electric currents created in sunward ionosphere.

Within approximately ± 20 degrees of the *magnetic equator*, is the *equatorial anomaly*. It is the occurrence of a trough of concentrated ionisation in the F_2 layer.

The Earth's magnetic field lines are horizontal at the magnetic equator. Solar heating and tidal oscillations in the lower ionosphere move plasma up and across the magnetic field lines. This sets up a sheet of electric current in the E region which, with the horizontal magnetic field, forces ionisation up into the F layer, concentrating at ± 20 degrees from the magnetic equator. This phenomenon is known as the *equatorial fountain*.

Equatorial Electrojet

The worldwide solar-driven wind results in the so-called Sq (solar quiet) current system in the E region of the Earth's ionosphere (100–130 km altitude). Resulting from this current is an electrostatic field directed E-W (dawn-dusk) in the equatorial day side of the ionosphere. At the magnetic dip equator, where the geomagnetic field is horizontal, this electric field results in an enhanced eastward current flow within ± 3 degrees of the magnetic equator, known as the equatorial electrojet.

Ionospheric Perturbations

X-rays: Sudden Ionospheric Disturbances (SID): When the Sun is active, strong solar flares can occur that will hit the Earth with hard X-rays on the sunlit side of the Earth. They will penetrate to the D-region, release electrons which will rapidly increase absorption causing a High Frequency (3 - 30 MHz) radio blackout. During this time Very Low Frequency (3 – 30 kHz) signals will become reflected by the D layer instead of the E layer, where the increased atmospheric density will usually increase the absorption of the wave, and thus dampen it. As soon as the X-rays end, the sudden ionospheric disturbance (SID) or radio black-out ends as the electrons in the D-region recombine rapidly and signal strengths return to normal.

Protons: Polar Cap Absorption (PCA)

Associated with solar flares is a release of high-energy protons. These particles can hit the Earth within 15 minutes to 2 hours of the solar flare. The protons spiral around and down the magnetic field lines of the Earth and penetrate into the atmosphere near the magnetic poles increasing the ionisation of the D and E layers. PCA's typically last anywhere from about an hour to several days, with an average of around 24 to 36 hours.

Geomagnetic Storms

A geomagnetic storm is a temporary intense disturbance of the Earth's magnetosphere.

- During a geomagnetic storm the F_2 layer will become unstable, fragment, and may even disappear completely.
- In the Northern and Southern pole regions of the Earth aurora will be observable in the sky.

Lightning

Lightning can cause ionospheric perturbations in the D-region in one of two ways. The first is through VLF frequency radio waves launched into the magnetosphere. These so-called "whistler" mode waves can interact with radiation belt particles and cause them to precipitate onto the ionosphere, adding ionisation to the D-region. These disturbances are called Lightning-induced Electron Precipitation (LEP) events.

Additional ionisation can also occur from direct heating/ionisation as a result of huge motions of charge in lightning strikes. These events are called Early/Fast.

In 1925, C. F. Wilson proposed a mechanism by which electrical discharge from lightning storms could propagate upwards from clouds to the ionosphere. Around the same time, Robert Watson-Watt, working at the Radio Research Station in Slough, UK, suggested that the ionospheric sporadic E layer (Es) appeared to be enhanced as a result of lightning but that more work was needed. In 2005, C. Davis and C. Johnson, working at the Rutherford Appleton Laboratory in Oxfordshire, UK, demonstrated that the Es layer was indeed enhanced as a result of lightning activity. Their subsequent research has focussed on the mechanism by which this process can occur.

Radio Application

DX communication, popular among amateur radio enthusiasts, is a term given to communication over great distances. Thanks to the property of ionised atmospheric gases to refract high frequency (HF, or shortwave) radio waves, the ionosphere can be utilised to "bounce" a transmitted signal down to ground. Transcontinental HF-connections rely on up to 5 bounces, or hops. Such communications played an important role during World War II. Karl Rawer's most sophisticated prediction method took account of several (zig-zag) paths, attenuation in the D-region and predicted the 11-years solar cycle by a method due to Wolfgang Gleißberg.

Mechanism of Refraction

When a radio wave reaches the ionosphere, the electric field in the wave forces the electrons in the ionosphere into oscillation at the same frequency as the radio wave. Some of the radio-frequency energy is given up to this resonant oscillation. The oscillating electrons will then either be lost to recombination or will re-radiate the original wave energy. Total refraction can occur when the collision frequency of the ionosphere is less than the radio frequency, and if the electron density in the ionosphere is great enough.

The critical frequency is the limiting frequency at or below which a radio wave is reflected by an ionospheric layer at vertical incidence. If the transmitted frequency is higher than the plasma frequency of the ionosphere, then the electrons cannot respond fast enough, and they are not able to re-radiate the signal. It is calculated as shown below:

$$f_{critical} = 9 \times 10^{-3}\sqrt{N}$$

where N = electron density per cm^3 and $f_{critical}$ is in MHz.

The Maximum Usable Frequency (MUF) is defined as the upper frequency limit that can be used for transmission between two points at a specified time.

$$f_{muf} = \frac{f_{critical}}{\sin\alpha}$$

where α = angle of attack, the angle of the wave relative to the horizon, and sin is the sine function.

The cutoff frequency is the frequency below which a radio wave fails to penetrate a layer of the ionosphere at the incidence angle required for transmission between two specified points by refraction from the layer.

Other Applications

The open system electrodynamic tether, which uses the ionosphere, is being researched. The space tether uses plasma contactors and the ionosphere as parts of a circuit to extract energy from the Earth's magnetic field by electromagnetic induction.

Measurements

Ionograms: Ionograms show the virtual heights and critical frequencies of the ionospheric layers and which are measured by an ionosonde. An ionosonde sweeps a range of frequencies, usually from 0.1 to 30 MHz, transmitting at vertical incidence to the ionosphere. As the frequency increases, each wave is refracted less by the ionisation in the layer, and so each penetrates further before it is reflected. Eventually, a frequency is reached that enables the wave to penetrate the layer without being reflected. For ordinary mode waves, this occurs when the transmitted frequency just exceeds the peak plasma, or critical, frequency of the layer. Tracings of the reflected high frequency radio pulses are known as ionograms. Reduction rules are given in: "URSI Handbook of Ionogram Interpretation and Reduction", edited by William Roy Piggott and Karl Rawer, Elsevier Amsterdam, 1961.

Incoherent Scatter Radars

Incoherent scatter radars operate above the critical frequencies. Therefore the technique allows to probe the ionosphere, unlike ionosondes, also above the electron density peaks. The thermal fluctuations of the electron density scattering the transmitted signals lack coherence, which gave the technique its name. Their power spectrum contains information not only on the density, but also on the ion and electron temperatures, ion masses and drift velocities.

Solar Flux

Solar flux is a measurement of the intensity of solar radio emissions at a frequency of 2800 MHz made using a radio telescope located in Dominion Radio Astrophysical Observatory, Penticton, British Columbia, Canada. Known also as the 10.7 cm flux (the wavelength of the radio signals at 2800 MHz), this solar radio emission has been shown to be proportional to sunspot activity. However, the level of the Sun's ultraviolet and X-ray emissions is primarily responsible for causing ionisation in the Earth's upper atmosphere. We now have data from the GOES spacecraft that measures the background X-ray flux from the Sun, a parameter more closely related to the ionisation levels in the ionosphere.

- The A and K indices are a measurement of the behaviour of the horizontal component of the geomagnetic field. The K index uses a scale from 0 to 9 to measure the change in the horizontal component of the geomagnetic field. A new K index is determined at the Table Mountain Observatory, north of Boulder, Colorado.
- The geomagnetic activity levels of the Earth are measured by the fluctuation of the Earth's magnetic field in SI units

called teslas (or in non-SI gauss, especially in older literature). The Earth's magnetic field is measured around the planet by many observatories. The data retrieved is processed and turned into measurement indices. Daily measurements for the entire planet are made available through an estimate of the *ap* index, called the *planetary A-index* (PAI).

Scientific Research on Ionospheric Propagation

Scientists also are exploring the structure of the ionosphere by a wide variety of methods, including passive observations of optical and radio emissions generated in the ionosphere, bouncing radio waves of different frequencies from it, incoherent scatter radars such as the EISCAT, Sondre Stromfjord, Millstone Hill, Arecibo, and Jicamarca radars, coherent scatter radars such as the Super Dual Auroral Radar Network (Super DARN) radars, and using special receivers to detect how the reflected waves have changed from the transmitted waves.

A variety of experiments, such as HAARP (High Frequency Active Auroral Research Program), involve high power radio transmitters to modify the properties of the ionosphere. These investigations focus on studying the properties and behaviour of ionospheric plasma, with particular emphasis on being able to understand and use it to enhance communications and surveillance systems for both civilian and military purposes. HAARP was started in 1993 as a proposed twenty year experiment, and is currently active near Gakona, Alaska.

The Super DARN radar project researches the high- and mid-latitudes using coherent backscatter of radio waves in the 8 to 20 MHz range. Coherent backscatter is similar to Bragg scattering in crystals and involves the constructive interference of scattering from ionospheric density irregularities. The project involves more than 11 different countries and multiple radars in both hemispheres.

Scientists are also examining the ionosphere by the changes to radio waves from satellites and stars passing through it. The Arecibo radio telescope located in Puerto Rico, was originally intended to study Earth's ionosphere.

Ionospheres on Other Planets and Titan

The atmosphere of Titan, the only moon known to have one, includes an ionosphere. It ranges from about 1100 to 1300 km in altitude, and contains carbon compounds.

Planets with ionospheres (incomplete list):

- Ionosphere of Venus
- Ionosphere of Uranus.

History

Guglielmo Marconi received the first trans-Atlantic radio signal on December 12, 1901, in St. John's, Newfoundland (now in Canada) using a 152.4 m (500 ft) kite-supported antenna for reception. The transmitting station in Poldhu, Cornwall used a spark-gap transmitter to produce a signal with a frequency of approximately 500 kHz and a power of 100 times more than any radio signal previously produced. The message received was three dits, the Morse code for the letter S. To reach Newfoundland the signal would have to bounce off the ionosphere twice. Dr. Jack Belrose has recently contested this, however, based on theoretical and experimental work. However, Marconi did achieve transatlantic wireless communications beyond a shadow of doubt in Glace Bay, Nova Scotia one year later.

In 1902, Oliver Heaviside proposed the existence of the *Kennelly-Heaviside Layer* of the ionosphere which bears his name. Heaviside's proposal included means by which radio signals are transmitted around the Earth's curvature. Heaviside's proposal, coupled with Planck's law of black body radiation, may have hampered the growth of radio astronomy for the detection of electromagnetic waves from celestial bodies until 1932 (and the development of high frequency radio transceivers). Also in 1902, Arthur Edwin Kennelly discovered some of the ionosphere's radio-electrical properties.

In 1912, the U.S. Congress imposed the Radio Act of 1912 on amateur radio operators, limiting their operations to frequencies above 1.5 MHz (wavelength 200 meters or smaller). The government thought those frequencies were useless. This led to the discovery of HF radio propagation via the ionosphere in 1923.

In 1926, Scottish physicist Robert Watson-Watt introduced the term *ionosphere* in a letter published only in 1969 in *Nature*:

We have in quite recent years seen the universal adoption of the term 'stratosphere'..and..the companion term 'troposphere'... The term 'ionosphere', for the region in which the main characteristic is large scale ionisation with considerable mean free paths, appears appropriate as an addition to this series.

Edward V. Appleton was awarded a Nobel Prize in 1947 for his confirmation in 1927 of the existence of the ionosphere. Lloyd

Berkner first measured the height and density of the ionosphere. This permitted the first complete theory of short wave radio propagation. Maurice V. Wilkes and J. A. Ratcliffere searched the topic of radio propagation of very long radio waves in the ionosphere. Vitaly Ginzburghas developed a theory of electromagnetic wave propagation in plasmas such as the ionosphere.

In 1962 the Canadian satellite Alouette 1 was launched to study the ionosphere. Following its success were Alouette 2 in 1965 and the two ISIS satellites in 1969 and 1971, further AEROS -A and -B in 1972 and 1975, all for measuring the ionosphere.

Aurora

An *aurora* (plural: *auroras* or *aurorae*) is a natural light display in the sky particularly in the high latitude (Arctic and Antarctic) regions, caused by the collision of energetic charged particles with atoms in the high altitude atmosphere (thermosphere). The charged particles originate in the magnetosphere and solar wind and are directed by the Earth's magnetic field into the atmosphere. Aurora is classified as diffuse or discrete aurora. Most aurorae occur in a band known as the *auroral zone* which is typically 3° to 6° in latitudinal extent and at all local times or longitudes. The auroral zone is typically 10° to 20° from the magnetic pole defined by the axis of the Earth's magnetic dipole. During a geomagnetic storm, the auroral zone will expand to lower latitudes. The diffuse aurora is a featureless glow in the sky which may not be visible to the naked eye even on a dark night and defines the extent of the auroral zone. The discrete aurora are sharply defined features within the diffuse aurora which vary in brightness from just barely visible to the naked eye to bright enough to read a newspaper at night. Discrete aurorae are usually observed only in the night sky because they are not as bright as the sunlit sky. Aurorae occur occasionally poleward of the auroral zone as diffuse patches or arcs (polar cap arcs) which are generally invisible to the naked eye.

In northern latitudes, the effect is known as the *aurora borealis* (or the northern lights), named after the Roman goddess of dawn, Aurora, and the Greek name for the north wind, Boreas, by Pierre Gassendi in 1621. Auroras seen near the magnetic pole may be high overhead, but from farther away, they illuminate the northern horizon as a greenish glow or sometimes a faint red, as if the Sun were rising from an unusual direction. Discrete aurorae often display magnetic field lines or curtain-like structures, and can change within seconds or glow unchanging for hours, most often in fluorescent green. The aurora borealis most often occurs near the equinoxes. The northern lights have had a number of names throughout history. The Cree call this phenomenon the "Dance of the Spirits". In Europe, in the Middle Ages, the auroras were commonly believed a sign from God.

Its southern counterpart, the *aurora australis* (or the southern lights), has almost identical features to the *aurora borealis* and changes simultaneously with changes in the northern auroral zone and is visible from high southern latitudes in Antarctica, South America and Australia.

Aurorae occur on other planets. Similar to the Earth's aurora, they are visible close to the planet's magnetic poles.

Auroral Mechanism

Auroras are result from emissions of photons in the Earth's upper atmosphere, above 80 km (50 mi), from ionised nitrogen atoms regaining an electron, and oxygen and nitrogen atoms returning from an excited state to ground state. They are ionised or excited by the collision of solar wind and magnetospheric particles being funnelled down and accelerated along the Earth's magnetic field lines; excitation energy is lost by the emission of a photon of light, or by collision with another atom or molecule:

- oxygen emissions
    - o Green or brownish-red, depending on the amount of energy absorbed.
- nitrogen emissions
    - o Blue or red. Blue if the atom regains an electron after it has been ionised. Red if returning to ground state from an excited state.

Oxygen is unusual in terms of its return to ground state: it can take three quarters of a second to emit green light and up to two minutes to emit red. Collisions with other atoms or molecules will absorb the excitation energy and prevent emission. Because the very top of the atmosphere has a higher percentage of oxygen and is sparsely distributed such collisions are rare enough to allow time for oxygen to emit red. Collisions become more frequent progressing down into the atmosphere, so that red emissions do not have time to happen, and eventually even green light emissions are prevented.

This is why there is a colour differential with altitude; at high altitude oxygen red dominates, then oxygen green and nitrogen blue/

red, then finally nitrogen blue/red when collisions prevent oxygen from emitting anything. Green is the most common of all auroras. Behind it is pink, a mixture of light green and red, followed by pure red, yellow (a mixture of red and green), and lastly pure blue.

Auroras are associated with the solar wind, a flow of ions continuously flowing outward from the Sun. The Earth's magnetic field traps these particles, many of which travel toward the poles where they are accelerated toward Earth. Collisions between these ions and atmospheric atoms and molecules cause energy releases in the form of auroras appearing in large circles around the poles. Auroras are more frequent and brighter during the intense phase of the solar cycle when coronal mass ejections increase the intensity of the solar wind.

Forms and Magnetism

Typically the aurora appears either as a diffuse glow or as "curtains" that approximately extend in the east-west direction. At some times, they form "quiet arcs"; at others ("active aurora"), they evolve and change constantly. Each curtain consists of many parallel rays, each lined up with the local direction of the magnetic field lines, suggesting that auroras are shaped by Earth's magnetic field. Indeed, satellites show electrons to be guided by magnetic field lines, spiraling around them while moving towards Earth.

The similarity to curtains is often enhanced by folds called "striations". When the field line guiding a bright auroral patch leads to a point directly above the observer, the aurora may appear as a "corona" of diverging rays, an effect of perspective.

Although it was first mentioned by Ancient Greekexplorer/ geographer Pytheas, Hiorter and Celsius first described in 1741 evidence for magnetic control, namely, large magnetic fluctuations occurred whenever the aurora was observed overhead. This indicates (it was later realised) that large electric currents were associated with the aurora, flowing in the region where auroral light originated. Kristian Birkeland (1908) deduced that the currents flowed in the east-west directions along the auroral arc, and such currents, flowing from the dayside towards (approximately) midnight were later named "auroral electrojets".

On 26 February 2008, THEMIS probes were able to determine, for the first time, the triggering event for the onset of magnetospheric substorms. Two of the five probes, positioned approximately one third the distance to the moon, measured events suggesting a magnetic reconnection event 96 seconds prior to auroral intensification. Dr. Vassilis Angelopoulos of the University of California, Los Angeles, the principal investigator for the THEMIS mission, claimed, "Our data show clearly and for the first time that magnetic reconnection is the trigger."

Still more evidence for a magnetic connection are the statistics of auroral observations. Elias Loomis(1860) and later in more detail Hermann Fritz (1881) and S. Tromholt (1882) established that the aurora appeared mainly in the "auroral zone", a ring-shaped region with a radius of approximately 2500 km around Earth's magnetic pole. It was hardly ever seen near the geographic pole, which is about 2000 km away from the magnetic pole. The instantaneous distribution of auroras ("auroral oval") is slightly different, centred about 3–5 degrees nightward of the magnetic pole, so that auroral arcs reach furthest towards the equator around midnight. The aurora can be seen best at this time.

Solar Wind and the Magnetosphere

The Earth is constantly immersed in the solar wind, a rarefied flow of hot plasma (gas of free electrons and positive ions) emitted by the Sun in all directions, a result of the two-million-degree heat of the Sun's outermost layer, the corona. The solar wind usually reaches Earth with a velocity around 400 km/s, density around 5 ions/cm^3 and magnetic field intensity around 2–5 nT (nanoteslas; Earth's surface field is typically 30,000–50,000 nT). These are typical values. During magnetic storms, in particular, flows can be several times faster; the interplanetary magnetic field (IMF) may also be much stronger.

The IMF originates on the Sun, related to the field of sunspots, and its field lines (lines of force) are dragged out by the solar wind. That alone would tend to line them up in the Sun-Earth direction, but the rotation of the Sun skews them (at Earth) by about 45 degrees, so that field lines passing Earth may actually start near the western edge ("limb") of the visible Sun.

Earth's magnetosphere is formed by the impact of the solar wind on the Earth's magnetic field. It forms an obstacle to the solar wind, diverting it, at an average distance of about 70,000 km (11 Earth radii or Re), forming a bow shock 12,000 km to 15,000 km (1.9 to 2.4 Re) further upstream. The width of the magnetosphere abreast of Earth, is typically 190,000 km (30 Re), and on the night side a long "magnetotail" of stretched field lines extends to great distances (> 200 Re).

The magnetosphere is full of trapped plasma as the solar wind passes the Earth. The flow of plasma into the magnetosphere increases with increases in solar wind density and speed, with increase in the southward component of the IMF and with increases in turbulence in the solar wind flow. The flow pattern of magnetospheric plasma is from the magnetotail toward the Earth, around the Earth and back into the solar wind through the magnetopause on the day-side. In addition to moving perpendicular to the Earth's magnetic field, some magnetospheric plasma travel down along the Earth's magnetic field lines and lose energy to the atmosphere in the auroral zones. Magnetospheric electrons which are accelerated downward by field-aligned electric fields are responsible for the bright aurora features. The un-accelerated electrons and ions are responsible for the dim glow of the diffuse aurora.

Frequency of Occurrence

Auroras are occasionally seen in temperate latitudes, when a magnetic storm temporarily grows the auroral oval. Large magnetic storms are most common during the peak of the eleven-year sunspot cycle or during the three years after that peak. However, within the auroral zone the likelihood of an aurora occurring depends mostly on the slant of interplanetary magnetic field (IMF) lines (the slant is known as B_z), being greater with southward slants.

Geomagnetic storms that ignite auroras actually happen more often during the months around the equinoxes. It is not well understood why geomagnetic storms are tied to Earth's seasons while polar activity is not. But it is known that during spring and autumn, the interplanetary magnetic field and that of Earth link up. At the magnetopause, Earth's magnetic field points north. When Bz becomes large and negative (i.e., the IMF tilts south), it can partially cancel Earth's magnetic field at the point of contact. South-pointing B_z's open a door through which energy from the solar wind can reach Earth's inner magnetosphere.

The peaking of B_z during this time is a result of geometry. The IMF comes from the Sun and is carried outward with the solar wind. The rotation of the Sun causes the IMF to have a spiral shape called the Parker spiral. The southward (and northward) excursions of B_z are greatest during April and October, when Earth's magnetic dipole axis is most closely aligned with the Parker spiral.

However, B_z is not the only influence on geomagnetic activity. The Sun's rotation axis is tilted 8 degrees with respect to the plane of Earth's orbit. The solar wind blows more rapidly from the Sun's poles than from its equator, thus the average speed of particles buffeting Earth's magnetosphere waxes and wanes every six months. The solar wind speed is greatest – by about 50 km/s, on average – around 5 September and 5 March when Earth lies at its highest heliographic latitude.

Still, neither B_z nor the solar wind can fully explain the seasonal behaviour of geomagnetic storms. Those factors together contribute only about one-third of the observed semiannual variations.

Auroral Events of Historical Significance

The auroras that resulted from the "great geomagnetic storm" on both 28 August and 2 September 1859 are thought the most spectacular in recent recorded history. Balfour Stewart, in a paper to the Royal Society on 21 November 1861, described both auroral events as documented by a self-recording magnetograph at the Kew Observatory and established the connection between the 2 September 1859 auroral storm and the Carrington-Hodgson flare event when he observed that "it is not impossible to suppose that in this case our luminary was taken *in the act*."

The second auroral event, which occurred on 2 September 1859 as a result of the exceptionally intense Carrington-Hodgson white light solar flare on 1 September 1859 produced auroras so widespread and extraordinarily brilliant that they were seen and reported in published scientific measurements, ships' logs and newspapers throughout the United States, Europe, Japan and Australia. It was reported by the *New York Times* that in Boston on Friday 2 September 1859 the aurora was "so brilliant that at about one o'clock ordinary print could be read by the light". One o'clock Boston time on Friday 2 September, would have been 6:00 GMT and the self-recording magnetograph at the Kew Observatory was recording the geomagnetic storm, which was then one hour old, at its full intensity.

Between 1859 and 1862, Elias Loomis published a series of nine papers on the Great Auroral Exhibition of 1859 in the *American Journal of Science* where he collected world wide reports of the auroral event. The aurora is thought to have been produced by one of the most intense coronal mass ejections in history, very near the maximum intensity that the Sun is thought to be capable of producing. It is also notable for the fact that it is the first time where the phenomena of auroral activity and electricity were unambiguously linked. This insight was made possible not only due to scientific magnetometre measurements of the era but also as a result of a significant portion of the 125,000 miles (201,000 km) of telegraph lines then in service being significantly disrupted for many hours throughout the storm. Some telegraph lines however seem to have been of the appropriate length and orientation to produce a sufficient geomagnetically induced current from the electromagnetic field to allow for continued communication with the telegraph operators' power supplies switched off. The following conversation occurred between two operators of the American Telegraph Line between Boston and Portland, Maine, on the night of 2 September 1859 and reported in the *Boston Traveller*:

Boston operator (to Portland operator): "Please cut off your battery [power source] entirely for fifteen minutes."

Portland operator: "Will do so. It is now disconnected."

Boston: "Mine is disconnected, and we are working with the auroral current. How do you receive my writing?"

Portland: "Better than with our batteries on. – Current comes and goes gradually."

Boston: "My current is very strong at times, and we can work better without the batteries, as the aurora seems to neutralise and augment our batteries alternately, making current too strong at times for our relay magnets. Suppose we work without batteries while we are affected by this trouble."

Portland: "Very well. Shall I go ahead with business?"

Boston: "Yes. Go ahead."

The conversation was carried on for around two hours using no battery power at all and working solely with the current induced by the aurora, and it was said that this was the first time on record that more than a word or two was transmitted in such manner.

The effect of the Aurora on the electric telegraph is generally to increase or diminish the electric current generated in working the wires. Sometimes it entirely neutralises them, so that, in effect, no fluid is discoverable in them . The aurora borealis seems to be composed of a mass of electric matter, resembling in every respect, that generated by the electric galvanic battery. The currents from it change coming on the wires, and then disappear: the mass of the aurora rolls from the horizon to the zenith.

Origin

The ultimate energy source of the aurora is the solar wind flowing past the Earth. The magnetosphere and solar wind consist of plasma (ionised gas), which conducts electricity. It is well known (since Michael Faraday's [1791 – 1867] work around 1830) that when an electrical conductor is placed within a magnetic field while relative motion occurs in a direction that the conductor cuts *across* (or is cut *by*), rather than *along*, the lines of the magnetic field, an electric current is said to be induced into that conductor and electrons will flow within it. The amount of current flow is dependent upon a) the rate of relative motion, b) the strength of the magnetic field, c) the number of conductors ganged together and d) the distance between the conductor and the magnetic field, while the *direction* of flow is dependent upon the direction of relative motion. Dynamosmake use of this basic process ("the dynamo effect"), any and all conductors, solid or otherwise are so affected including plasmas or other fluids.

In particular the solar wind and the magnetosphere are two electrically conducting fluids with such relative motion and should be able (in principle) to generate electric currents by "dynamo action", in the process also extracting energy from the flow of the solar wind. The process is hampered by the fact that plasmas conduct easily along magnetic field lines, but not so easily perpendicular to them. So it is important that a temporary magnetic connection be established between the field lines of the solar wind and those of the magnetosphere, by a process known as magnetic reconnection. It happens most easily with a southward slant of interplanetary field lines, because then field lines north of Earth approximately match the direction of field lines near the north magnetic pole (namely, *into* Earth), and similarly near the south magnetic pole. Indeed, active auroras (and related "substorms") are much more likely at such times. Electric currents originating in such way apparently give auroral electrons their energy. The magnetospheric plasma has an abundance of electrons: some are magnetically trapped, some reside in the magnetotail, and some exist in the upward extension of the ionosphere, which may extend (with diminishing density) some 25,000 km around Earth.

Bright auroras are generally associated with Birkeland currents (Schield et al., 1969; Zmuda and Armstrong, 1973) which flow down into the ionosphere on one side of the pole and out on the other. In between, some of the current connects directly through the ionospheric E layer (125 km); the rest ("region 2") detours, leaving again through field lines closer to the equator and closing through the "partial ring current" carried by magnetically trapped plasma. The ionosphere is an ohmic conductor, so such currents require a driving voltage, which some dynamo mechanism can supply. Electric field probes in orbit above the polar cap suggest voltages of the order of 40,000 volts, rising up to more than 200,000 volts during intense magnetic storms.

Ionospheric resistance has a complex nature, and leads to a secondary Hall current flow. By a strange twist of physics, the magnetic disturbance on the ground due to the main current almost cancels out, so most of the observed effect of auroras is due to a secondary current, the auroral electrojet. An auroral electrojet index (measured in nanotesla) is regularly derived from ground data and serves as a general measure of auroral activity.

However, ohmic resistance is not the only obstacle to current flow in this circuit. The convergence of magnetic field lines near Earth creates a "mirror effect" that turns back most of the down-flowing electrons (where currents flow upwards), inhibiting current-carrying capacity. To overcome this, part of the available voltage appears along the field line ("parallel to the field"), helping electrons overcome that obstacle by widening the bundle of trajectories reaching Earth; a similar "parallel potential" is used in "tandem mirror" plasma containment devices. A feature of such voltage is that it is concentrated near Earth (potential proportional to field intensity; Persson, 1963), and indeed, as deduced by Evans (1974) and confirmed by satellites, most auroral acceleration occurs below 10,000 km. Another indicator of parallel electric fields along field lines are beams of upwards flowing O+ ions observed on auroral field lines.

Some O+ ions ("conics") also seem accelerated in different ways by plasma processes associated with the aurora. These ions are accelerated by plasma waves, in directions mainly perpendicular to the field lines. They therefore start at their own "mirror points" and can travel only upwards. As they do so, the "mirror effect" transforms their directions of motion, from perpendicular to the line to lying on a cone around it, which gradually narrows down.

In addition, the aurora and associated currents produce a strong radio emission around 150 kHz known as auroral kilometric radiation (AKR, discovered in 1972). Ionospheric absorption makes AKR observable from space only.

These "parallel potentials" accelerate electrons to auroral energies and seem to be a major source of aurora. Other mechanisms have also been proposed, in particular, Alfvén waves, wave modes involving the magnetic field first noted by Hannes Alfvén (1942), which have been observed in the lab and in space. The question is however whether these waves might just be a different way of looking at the above process, because this approach does not point out a different energy source, and many plasma bulk phenomena can also be described in terms of Alfvén waves.. There are three possible main sources:

1. Dynamo action with the solar wind *flowing past Earth*, possibly producing quiet auroral arcs ("directly driven" process). The circuit of the accelerating currents and their connection to the solar wind are uncertain.
2. Dynamo action involving plasma squeezed towards Earth by sudden convulsions of the magnetotail ("magnetic substorms"). Substorms tend to occur after prolonged spells (hours) during which the interplanetary magnetic field has an appreciable southward component, leading to a high rate of interconnection between its field lines and those of Earth. As a result the solar wind moves magnetic flux (tubes of magnetic field lines, moving together with their resident plasma) from the day side of Earth to the magnetotail, widening the obstacle it presents to the solar wind flow and causing it to be squeezed harder. Ultimately the tail plasma is torn ("magnetic reconnection"); some blobs ("plasmoids") are squeezed tailwards and are carried away with the solar wind; others are squeezed towards Earth where their motion feeds large outbursts of aurora, mainly around midnight ("unloading process"). Geomagnetic storms have similar effects, but with greater vigour. The big difference is the addition of many particles to the plasma trapped around Earth, enhancing the "ring current" it carries. The resulting modification of Earth's field makes auroras visible at middle latitudes, on field lines much closer to the equator.
3. Satellite images of the aurora from above show a "ring of fire" along the auroral oval, often widest at midnight. That is the "diffuse aurora", not distinct enough to be seen by the eye. It does *not* seem to be associated with acceleration by electric currents (although currents and their arcs may be embedded in it) but to be due to electrons leaking out of the magnetotail.

Any magnetic trapping is leaky—there always exists a bundle of directions ("loss cone") around the guiding magnetic field lines where particles are not trapped but escape. In the radiation belts of Earth, once particles on such trajectories are gone, new ones only replace them very slowly, leaving such directions nearly "empty". In the magnetotail, however, particle trajectories seem to be constantly reshuffled, probably when the particles cross the very weak field near the equator. As a result, the flow of electrons in all directions is nearly the same ("isotropic"), and that assures a steady supply of leaking electrons.

The energisation of such electrons comes from magnetotail processes. The leakage of negative electrons does not leave the tail positively charged, because each leaked electron lost to the atmosphere is quickly replaced by a low energy electron drawn upwards from the ionosphere. Such replacement of "hot" electrons by "cold" ones is in complete accord with the 2nd law of thermodynamics.

Other types of auroras have been observed from space, e.g. "poleward arcs" stretching sunward across the polar cap, the related "theta aurora", and "dayside arcs" near noon. These are relatively infrequent and poorly understood. There are other interesting effects such as flickering aurora, "black aurora" and subvisual red arcs. In addition to all these, a weak glow (often deep red) has been observed around the two polar cusps, the "funnels" of field lines separating the ones that close on the day side of Earth from lines swept into the tail. The cusps allow a small amount of solar wind to reach the top of the atmosphere, producing an auroral glow.

On Other Planets

Both Jupiter and Saturn have magnetic fields much stronger than Earth's (Jupiter's equatorial field strength is 4.3 gauss, compared to 0.3 gauss for Earth), and both have large radiation belts. Auroras have been observed on both, most clearly with the Hubble Space Telescope. Uranus andNeptune have also been observed to have auroras.

The auroras on the gas giants seem, like Earth's, to be powered by the solar wind. In addition, however, Jupiter's moons, especially Io, are powerful sources of auroras on Jupiter. These arise from electric currents along field lines ("field aligned currents"), generated by a dynamo mechanism due to the relative motion between the rotating planet and the moving moon. Io, which has active volcanism and an ionosphere, is a particularly strong source, and its currents also generate radio emissions, studied since 1955. Auroras have also been observed on Io, Europa, and Ganymede themselves, e.g., using the Hubble Space Telescope. These Auroras have also been observed on Venus and Mars. Because Venus has no intrinsic (planetary) magnetic field, Venusian auroras appear as bright and diffuse patches of varying shape and intensity, sometimes distributed across the full planetary disc. Venusian auroras are produced by the impact of electrons originating from the solar wind and precipitating in the night-side atmosphere. An aurora was also detected on Mars, on 14 August 2004, by the SPICAM instrument aboard Mars Express.

The aurora was located at Terra Cimmeria, in the region of 177° East, 52° South. The total size of the emission region was about 30 km across, and possibly about 8 km high.

By analysing a map of crustal magnetic anomalies compiled with data from Mars Global Surveyor, scientists observed that the region of the emissions corresponded to an area where the strongest magnetic field is localised. This correlation indicates that the origin of the light emission was a flux of electrons moving along the crust magnetic lines and exciting the upper atmosphere of Mars.

History of Aurora Theories

In the past theories have been proposed to explain the phenomenon. These theories are now obsolete.

- Seneca speaks diffusely on auroras in the first book of his Naturales Quaestiones, drawing mainly from Aristoteles; he classifies them ("putei" or wells when they are circular and "rim a large hole in the sky", "pithaei" when they look like casks, "chasmata" from the same root of the English chasm, "pogoniae" when they are bearded, "cyparissae" when they look like cypresses), describes their manifold colours and asks himself whether they are above or below the clouds. He recalls that under Tiberius, an aurora formed above Ostia, so intense and so red that a cohort of the army, stationed nearby for fireman duty, galloped to the city.
- Benjamin Franklin theorised that the "mystery of the Northern Lights" was caused by a concentration of electrical charges in the polar regions intensified by the snow and other moisture.
- Auroral electrons come from beams emitted by the Sun. This was claimed around 1900 by Kristian Birkeland, whose experiments in a vacuum chamber with electron beams and magnetised spheres (miniature models of Earth or "terrellas") showed that such electrons would be guided towards the polar regions. Problems with this model included absence of aurora at the poles themselves, self-dispersal of such beams by their

negative charge, and more recently, lack of any observational evidence in space.

- The aurora is the overflow of the radiation belt ("leaky bucket theory"). This was first disproved around 1962 by James Van Allen and co-workers, who showed that the high rate of energy dissipation by the aurora would quickly drain the radiation belt. Soon afterward, it became clear that most of the energy in trapped particles resided in positive ions, while auroral particles were almost always electrons, of relatively low energy.
- The aurora is produced by solar wind particles guided by Earth's field lines to the top of the atmosphere. This holds true for the cusp aurora, but outside the cusp, the solar wind has no direct access. In addition, the main energy in the solar wind resides in positive ions; electrons only have about 0.5 eV (electron volt), and while in the cusp this may be raised to 50–100 eV, that still falls short of auroral energies.
- After the Battle of Fredericksburg the lights could be seen from the battlefield that night. The Confederate army took it as a sign that God was on their side during the battle. It was very rare that one could see the Lights in Virginia.

Images

Images of auroras are significantly more common today due to the rise of use of digital cameras that have high enough sensitivities. Film and digital exposure to auroral displays is fraught with difficulties, particularly if faithfulness of reproduction is an objective. Due to the different spectral energy present, and changing dynamically throughout the exposure, the results are somewhat unpredictable. Different layers of the film emulsion respond differently to lower light levels, and choice of film can be very important. Longer exposures aggregate the rapidly changing energy and often blanket the dynamic attribute of a display. Higher sensitivity creates issues with graininess.

David Malin pioneered multiple exposure using multiple filters for astronomical photography, recombining the images in the laboratory to recreate the visual display more accurately. For scientific research, proxies are often used, such as ultra-violet, and re-coloured to simulate the appearance to humans. Predictive techniques are also used, to indicate the extent of the display, a highly useful tool for aurora hunters. Terrestrial features often find their way into aurora images, making them more accessible and more likely to be published by the major websites. It is possible to take excellent images with standard film (using ISO ratings between 100 and 400) and a single-lens reflex camera with full aperture, a fast lens (f1.4 50 mm, for example), and exposures between 10 and 30 seconds, depending on the aurora's display strength.

Early work on the imaging of the auroras was done in 1949 by the University of Saskatchewan using the SCR-270 radar.

In Traditional and Popular Culture

In *Bulfinch's Mythology* from 1855 by Thomas Bulfinch there is the claim that in Norse mythology:

The Valkyrior are warlike virgins, mounted upon horses and armed with helmets and spears. /.../ When they ride forth on their errand, their armour sheds a strange flickering light, which flashes up over the northern skies, making what men call the "aurora borealis", or "Northern Lights".

While a striking notion, there is not a vast body of evidence in the Old Norse literature supporting this assertion. Although auroral activity is common over Scandinavia and Iceland today, it is possible that the Magnetic North Pole was considerably further away from this region during the centuries before the documentation of Norse mythology, thus explaining the lack of references.

The first Old Norse account of *norðrljós* is found in the Norwegian chronicle *Konungs Skuggsjá* from AD 1230. The chronicler has heard about this phenomenon from compatriots returning from Greenland, and he gives three possible explanations: that the ocean was surrounded by vast fires, that the sun flares could reach around the world to its night side, or that glaciers could store energy so that they eventually became fluorescent.

In ancient Roman mythology, Aurora is the goddess of the dawn, renewing herself every morning to fly across the sky, announcing the arrival of the sun. The persona of Aurora the goddess has been incorporated in the writings of Shakespeare, Lord Tennyson and Thoreau.

Chapter 5

Kinetics

In physics and engineering, kinetics is a term for the branch of classical mechanics that is concerned with the relationship between the motion of bodies and its causes, namely forces and torques. Since the mid-20th century, the term "dynamics" (or "analytical dynamics") has largely superseded "kinetics" in physics text books; the term "kinetics" is still used in engineering.

The Kinetics is deduced from Kinematics by the introduction of the concept of mass. In plasma physics, kinetics refers to the study of continua in velocity space. This is usually in the context of non-thermal (non-Maxwellian) velocity distributions, or processes that perturb thermal distributions. These plasmas cannot be adequately described with fluid equations.

Such plasmas are termed kinetic plasmas. The term *kinetics* is also used to refer to chemical kinetics, particularly in chemical physics and physical chemistry. In such uses, a qualifier is often used or implied.

Classical Mechanics

Classical mechanics is one of the two major sub-fields of mechanics, which is concerned with the set of physical laws describing the motion of bodies under the action of a system of forces. The study of the motion of bodies is an ancient one, making classical mechanics one of the oldest and largest subjects in science, engineering and technology.

Classical mechanics describes the motion of macroscopic objects, from projectiles to parts of machinery, as well as astronomical objects, such as spacecraft, planets, stars, and galaxies. Besides this, many specialisations within the subject deal with gases, liquids, solids, and other specific sub-topics. Classical mechanics provides extremely accurate results as long as the domain of study is restricted to large objects and the speeds involved do not approach the speed of light. When the objects being dealt with become sufficiently small, it becomes necessary to introduce the other major sub-field of mechanics, quantum mechanics, which reconciles the macroscopic laws of physics with the atomic nature of matter and handles the wave-particle duality of atoms and molecules. In the case of high velocity objects approaching the speed of light, classical mechanics is enhanced by special relativity. General relativity unifies special relativity with Newton's law of universal gravitation, allowing physicists to handle gravitation at a deeper level.

The term *classical mechanics* was coined in the early 20th century to describe the system of physics begun by Isaac Newton and many contemporary 17th century natural philosophers, building upon the earlier astronomical theories of Johannes Kepler, which in turn were based on the precise observations of Tycho Brahe and the studies of terrestrial projectile motion of Galileo. Because these aspects of physics were developed long before the emergence of quantum physics and relativity, some sources exclude Einstein's theory of relativity from this category. However, a number of modern sources *do* include relativistic mechanics, which in their view represents *classical mechanics* in its most developed and most accurate form.

The initial stage in the development of classical mechanics is often referred to as Newtonian mechanics, and is associated with the physical concepts employed by and the mathematical methods invented by Newton himself, in parallel with Leibniz, and others. This is further described in the following sections. Later, more abstract and general methods were developed, leading to reformulations of classical mechanics known as Lagrangian mechanics and Hamiltonian mechanics. These advances were largely made in the 18th and 19th centuries, and they extend substantially beyond Newton's work, particularly through their use of analytical mechanics. Ultimately, the mathematics developed for these were central to the creation of quantum mechanics.

Description of the Theory

The following introduces the basic concepts of classical mechanics. For simplicity, it often models real-world objects as point particles, objects with negligible size. The motion of a point particle is characterised by a small number of parameters: its position, mass, and the forces applied to it. Each of these parameters is discussed in turn. In reality, the kind of objects that classical mechanics can describe always have a non-zero size. (The physics of *very small* particles, such as the electron, is more accurately described by quantum mechanics). Objects with non-zero size have more complicated behaviour than hypothetical point particles, because of the additional degrees of freedom—for example, a baseball can spin while it is moving.

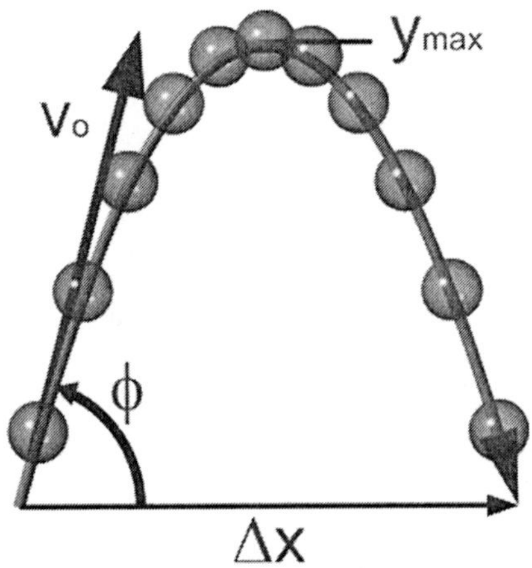

Figure 1: The analysis of projectile motion is a part of classical mechanics.

However, the results for point particles can be used to study such objects by treating them as composite objects, made up of a large number of interacting point particles. The centre of mass of a composite object behaves like a point particle.

Position and its Derivatives

Table 1: The SI derived "mechanical" (that is, not electromagnetic or thermal) units with kg, m and s

Position	m
Angular position/Angle	unitless (radian)
velocity	$m\ s^{-1}$
Angular velocity	s^{-1}
acceleration	$m\ s^{-2}$
Angular acceleration	s^{-2}
jerk	$m\ s^{-3}$
"Angular jerk"	s^{-3}
specific energy	$m^2\ s^{-2}$
absorbed dose rate	$m^2\ s^{-3}$

moment of inertia	kg m^2
momentum	kg m s^{-1}
angular momentum	kg m^2 s^{-1}
force	kg m s^{-2}
torque	kg m^2 s^{-2}
energy	kg m^2 s^{-2}
power	kg m^2 s^{-3}
pressure and energy density	kg m^{-1} s^{-2}
surface tension	kg s^{-2}
Spring constant	kg s^{-2}
irradiance and energy flux	kg s^{-3}
kinematic viscosity	m^2 s^{-1}
dynamic viscosity	kg m^{-1} s^{-1}
Density (mass density)	kg m^{-3}
Density (weight density)	kg m^{-2} s^{-2}
Number density	m^{-3}
Action	kg m^2 s^{-1}

The *position* of a point particle is defined with respect to an arbitrary fixed reference point, O, in space, usually accompanied by a coordinate system, with the reference point located at the *origin* of the coordinate system. It is defined as the vector r from Oto the particle. In general, the point particle need not be stationary relative to O, so r is a function of *t*, the time elapsed since an arbitrary initial time. In pre-Einstein relativity (known as Galilean relativity), time is considered an absolute, i.e., the time interval between any given pair of events is the same for all observers. In addition to relying on absolute time, classical mechanics assumes Euclidean geometry for the structure of space.

Velocity and Speed

The *velocity*, or the rate of change of position with time, is defined as the derivative of the position with respect to time or

$$v = \frac{dr}{dt} .$$

In classical mechanics, velocities are directly additive and subtractive. For example, if one car travelling East at 60 km/h passes another car travelling East at 50 km/h, then from the perspective of the slower car, the faster car is travelling east at 60 - 50 = 10 km/h. Whereas, from the perspective of the faster car, the slower car is moving 10 km/h to the West. Velocities are directly additive as vector quantities; they must be dealt with using vector analysis.

Mathematically, if the velocity of the first object in the previous discussion is denoted by the vector $u = u$d and the velocity of the second object by the vector $v = v$e, where u is the speed of the first object, v is the speed of the second object, and d and e are unit vectors in the directions of motion of each particle respectively, then the velocity of the first object as seen by the second object is

$$u' = u - v.$$

Similarly,

$$v' = v - u.$$

When both objects are moving in the same direction, this equation can be simplified to

$$u' = (u - v)d.$$

Or, by ignoring direction, the difference can be given in terms of speed only:

Acceleration

The *acceleration*, or rate of change of velocity, is the derivative of the velocity with respect to time (the second derivative of the position with respect to time) or

Acceleration can arise from a change with time of the magnitude of the velocity or of the direction of the velocity or both. If only the magnitude v of the velocity decreases, this is sometimes referred to as deceleration, but generally any change in the velocity with time, including deceleration, is simply referred to as acceleration.

Frames of Refesrence

While the position and velocity and acceleration of a particle can be referred to any observer in any state of motion, classical mechanics assumes the existence of a special family of reference frames in terms of which the mechanical laws of nature take a comparatively simple form. These special reference frames are called inertial frames. An inertial frame is such that when an object without any force interactions (an idealised situation) is viewed from it, it will appear either to be at rest or in a state of uniform motion in a straight line. This is the fundamental definition of an inertial frame. They are characterised by the requirement that all forces entering the observer's physical laws originate in identifiable sources (charges, gravitational bodies, and so forth). A non-inertial reference frame is one accelerating with respect to an inertial one, and in such a non-inertial frame a particle is subject to acceleration by fictitious forces that enter the equations of motion solely as a result of its accelerated motion, and do not originate in identifiable sources. These fictitious forces are in addition to the real forces recognised in an inertial frame. A key concept of inertial frames is the method for identifying them. For practical purposes, reference frames that are unaccelerated with respect to the distant stars are regarded as good approximations to inertial frames.

Consider two reference frames S and S'. For observers in each of the reference frames an event has space-time coordinates of (x,y,z,t) in frame S and $(x2, y2, z2, t2)$ in frame $S2$. Assuming time is measured the same in all reference frames, and if we require $x = x'$ when $t = 0$, then the relation between the space-time coordinates of the same event observed from the reference frames $S2$ and S, which are moving at a relative velocity of u in the x direction is:

$$x2 = x - ut$$

$$y2 = y$$

$$z2 = z$$

$$t2 = t$$

This set of formulas defines a group transformation known as the Galilean transformation (informally, the *Galilean transform*). This group is a limiting case of the Poincaré group used in special relativity. The limiting case applies when the velocity u is very small compared to c, the speed of light.

The transformations have the following consequences:

- v2 = v - u (the velocity v2 of a particle from the perspective of $S2$ is slower by u than its velocity v from the perspective of S)
- a2 = a (the acceleration of a particle is the same in any inertial reference frame)
- F2 = F (the force on a particle is the same in any inertial reference frame)
- the speed of light is not a constant in classical mechanics, nor does the special position given to the speed of light in relativistic mechanics have a counterpart in classical mechanics.

For some problems, it is convenient to use rotating coordinates (reference frames). Thereby one can either keep a mapping to a convenient inertial frame, or introduce additionally a fictitious centrifugal force and Coriolis force.

Forces; Newton's Second Law

Newton was the first to mathematically express the relationship between force and momentum. Some physicists interpret Newton's second law of motion as a definition of force and mass, while others consider it to be a fundamental postulate, a law of nature. Either interpretation has the same mathematical consequences, historically known as "Newton's Second Law":

$$F = \frac{dp}{dt} = \frac{d(mv)}{dt}.$$

The quantity *m*v is called the (canonical) momentum. The net force on a particle is thus equal to rate change of momentum of the particle with time. Since the definition of acceleration is a = dv/d*t*, the second law can be written in the simplified and more familiar form:

$$F = ma.$$

So long as the force acting on a particle is known, Newton's second law is sufficient to describe the motion of a particle. Once independent relations for each force acting on a particle are available, they can be substituted into Newton's second law to obtain an ordinary differential equation, which is called the *equation of motion.*

As an example, assume that friction is the only force acting on the particle, and that it may be modelled as a function of the velocity of the particle, for example:

$$F_R = -\lambda v,$$

where *ë* is a positive constant. Then the equation of motion is

$$-\lambda v = ma = m\frac{dv}{dt}.$$

This can be integrated to obtain

$$v = v_0 e^{-\lambda t/m}$$

where v_0 is the initial velocity. This means that the velocity of this particle decays exponentially to zero as time progresses. In this case, an equivalent viewpoint is that the kinetic energy of the particle is absorbed by friction (which converts it to heat energy in accordance with the conservation of energy), slowing it down. This expression can be further integrated to obtain the position r of the particle as a function of time. Important forces include the gravitational force and the Lorentz force for electromagnetism. In addition, Newton's third law can sometimes be used to deduce the forces acting on a particle: if it is known that particle A exerts a force F on another particle B, it follows that B must exert an equal and opposite reaction *force*, -F, on A. The strong form of Newton's third law requires that F and -F act along the line connecting A and B, while the weak form does not. Illustrations of the weak form of Newton's third law are often found for magnetic forces.

Work and Energy

If a constant force F is applied to a particle that achieves a displacement Δr, the *work done* by the force is defined as the scalar product of the force and displacement vectors:

$$W = F \cdot \Delta r.$$

More generally, if the force varies as a function of position as the particle moves from r_1 to r_2 along a path *C*, the work done on the particle is given by the line integral

$$W = \int_C F(r) \cdot dr.$$

If the work done in moving the particle from r_1 to r_2 is the same no matter what path is taken, the force is said to be conservative. Gravity is a conservative force, as is the force due to an idealised spring, as given by Hooke's law. The force due to friction is non-conservative.

The kinetic energy E_k of a particle of mass m travelling at speed v is given by

$$E_k = \tfrac{1}{2} m v^2.$$

For extended objects composed of many particles, the kinetic energy of the composite body is the sum of the kinetic energies of the particles.

The work-energy theorem states that for a particle of constant mass m the total work W done on the particle from position r_1 to r_2 is equal to the change in kinetic energy E_k of the particle:

$$W = \Delta E_k = E_{k,2} - E_{k,1} = \tfrac{1}{2} m \left(v_2^2 - v_1^2 \right).$$

Conservative forces can be expressed as the gradient of a scalar function, known as the potential energy and denoted E_p:

$$F = -\nabla E_p.$$

If all the forces acting on a particle are conservative, and E_p is the total potential energy (which is defined as a work of involved forces to rearrange mutual positions of bodies), obtained by summing the potential energies corresponding to each force

$$F \cdot \Delta r = -\nabla E_p \cdot \Delta r = -\Delta E_p \Rightarrow -\Delta E_p = \Delta E_k \Rightarrow \Delta(E_k + E_p) = 0.$$

This result is known as *conservation of energy* and states that the total energy,

$$\sum E = E_k + E_p.$$

is constant in time. It is often useful, because many commonly encountered forces are conservative.

Beyond Newton's Laws

Classical mechanics also includes descriptions of the complex motions of extended non-pointlike objects. Euler's laws provide extensions to Newton's laws in this area. The concepts of angular momentum rely on the same calculus used to describe one-dimensional motion. The rocket equation extends the notion of rate of change of an object's momentum to include the effects of an object "losing mass".

There are two important alternative formulations of classical mechanics: Lagrangian mechanics and Hamiltonian mechanics. These, and other modern formulations, usually bypass the concept of "force", instead referring to other physical quantities, such as energy, for describing mechanical systems.

The expressions given above for momentum and kinetic energy are only valid when there is no significant electromagnetic contribution. In electromagnetism, Newton's second law for current-carrying wires breaks down unless one includes the electromagnetic field contribution to the momentum of the system as expressed by the Poynting vector divided by c^2, where c is the speed of light in free space.

History

Some Greek philosophers of antiquity, among them Aristotle, founder of Aristotelian physics, may have been the first to maintain the idea that "everything happens for a reason" and that theoretical principles can assist in the understanding of nature. While to a modern reader, many of these preserved ideas come forth as eminently reasonable, there is a conspicuous lack of both mathematical theory and controlled experiment, as we know it. These both turned out to be decisive factors in forming modern science, and they started out with classical mechanics.

The medieval "science of weights" (i.e., mechanics) owes much of its importance to the work of Jordanus de Nemore. In the *Elementa super demonstrationem ponderum*, he introduces the concept of "positional gravity" and the use of component forces.

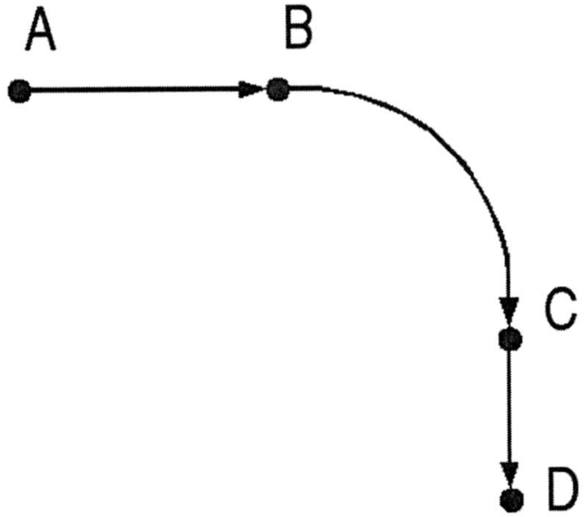

Figure 2: Three stage Theory of impetus according to Albert of Saxony.

The first published causal explanation of the motions of planets was Johannes Kepler's Astronomia nova published in 1609. He concluded, based on Tycho Brahe's observations of the orbit of Mars, that the orbits were ellipses. This break with ancient thought was happening around the same time that Galilei was proposing abstract mathematical laws for the motion of objects. He may (or may not) have performed the famous experiment of dropping two cannon balls of different weights from the tower of Pisa, showing that they both hit the ground at the same time. The reality of this experiment is disputed, but, more importantly, he did carry out quantitative experiments by rolling balls on an inclined plane. His theory of accelerated motion derived from the results of such experiments, and forms a cornerstone of classical mechanics.

As foundation for his principles of natural philosophy, Newton proposed three laws of motion: the law of inertia, his second law of acceleration (mentioned above), and the law of action and reaction; and hence laid the foundations for classical mechanics. Both Newton's second and third laws were given proper scientific and mathematical treatment in Newton's Philosophiæ Naturalis Principia Mathematica, which distinguishes them from earlier attempts at explaining similar phenomena, which were either incomplete, incorrect, or given little accurate mathematical expression. Newton also enunciated the principles of conservation of momentum and angular momentum. In Mechanics, Newton was also the first to provide the first correct scientific and mathematical formulation of gravity in Newton's law of universal gravitation. The combination of Newton's laws of motion and gravitation provide the fullest and most accurate description of classical mechanics. He demonstrated that these laws apply to everyday objects as well as to celestial objects. In particular, he obtained a theoretical explanation of Kepler's laws of motion of the planets.

Newton previously invented the calculus, of mathematics, and used it to perform the mathematical calculations. For acceptability, his book, the Principia, was formulated entirely in terms of the long established geometric methods, which were soon to be eclipsed by his calculus. However it was Leibniz who developed the notation of the derivative and integral preferred today.

Newton, and most of his contemporaries, with the notable exception of Huygens, worked on the assumption that classical mechanics would be able to explain all phenomena, including light, in the form of geometric optics. Even when discovering the so-called Newton's rings (a wave interference phenomenon) his explanation remained with his own corpuscular theory of light.

After Newton, classical mechanics became a principal field of study in mathematics as well as physics. After Newton there were several re-formulations which progressively allowed a solution to be found to a far greater number of problems. The first notable re-formulation was in 1788 by Joseph Louis Lagrange. Lagrangian mechanics was in turn re-formulated in 1833 by William Rowan Hamilton.

Some difficulties were discovered in the late 19th century that could only be resolved by more modern physics. Some of these difficulties related to compatibility with electromagnetic theory, and the famous Michelson-Morley experiment. The resolution of these problems led to the special theory of relativity, often included in the term classical mechanics.

A second set of difficulties were related to thermodynamics. When combined with thermodynamics, classical mechanics leads to the Gibbs paradox of classical statistical mechanics, in which entropy is not a well-defined quantity. Black-body radiation was not explained without the introduction of quanta. As experiments reached the atomic level, classical mechanics failed to explain, even approximately, such basic things as the energy levels and sizes of atoms and the photo-electric effect. The effort at resolving these problems led to the development of quantum mechanics.

Since the end of the 20th century, the place of classical mechanics in physics has been no longer that of an independent theory. Emphasis has shifted to understanding the fundamental forces of nature as in the Standard model and its more modern extensions into a unified theory of everything. Classical mechanics is a theory for the study of the motion of non-quantum mechanical, low-energy particles in weak gravitational fields.

In the 21st century classical mechanics has been extended into the complex domain and complex classical mechanics exhibits behaviours very similar to quantum mechanics.

Limits of Validity

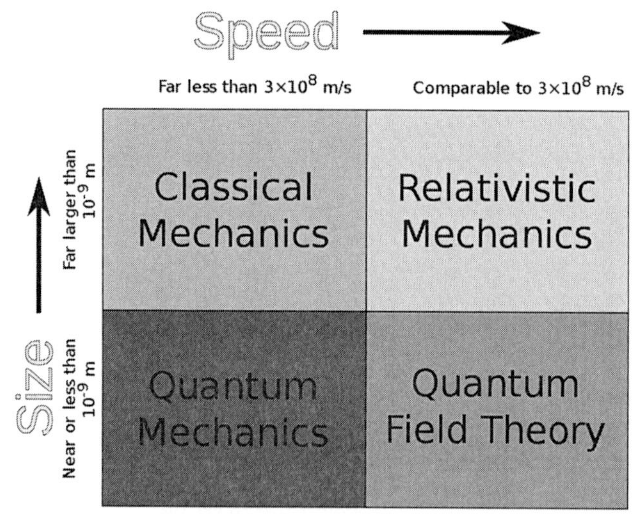

Figure 3: Domain of validity for Classical Mechanics

Many branches of classical mechanics are simplifications or approximations of more accurate forms; two of the most accurate being general relativity and relativistic statistical mechanics. Geometric optics is an approximation to the quantum theory of light, and does not have a superior "classical" form.

The Newtonian Approximation to Special Relativity

In special relativity, the momentum of a particle is given by

$$p = \frac{mv}{\sqrt{1-(v^2/c^2)}},$$

where m is the particle's mass, v its velocity, and c is the speed of light. If v is very small compared to c, v^2/c^2 is approximately zero, and so

$$p \approx mv.$$

Thus the Newtonian equation p = mv is an approximation of the relativistic equation for bodies moving with low speeds compared to the speed of light.

For example, the relativistic cyclotron frequency of a cyclotron, gyrotron, or high voltage magnetron is given by

$$f = f_c \frac{m_0}{m_0 + T / c^2},$$

where f_c is the classical frequency of an electron (or other charged particle) with kinetic energy T and (rest) mass m_0 circling in a magnetic field. The (rest) mass of an electron is 511 keV. So the frequency correction is 1% for a magnetic vacuum tube with a 5.11 kV direct current accelerating voltage.

The Classical Approximation to Quantum Mechanics

The ray approximation of classical mechanics breaks down when the de Broglie wavelength is not much smaller than other dimensions of the system. For non-relativistic particles, this wavelength is

$$\lambda = \frac{h}{p}$$

where h is Planck's constant and p is the momentum.

Again, this happens with electrons before it happens with heavier particles. For example, the electrons used by Clinton Davisson and Lester Germer in 1927, accelerated by 54 volts, had a wave length of 0.167 nm, which was long enough to exhibit a single diffraction side lobe when reflecting from the face of a nickel crystal with atomic spacing of 0.215 nm. With a larger vacuum chamber, it would seem relatively easy to increase the angular resolution from around a radian to a milliradian and see quantum diffraction from the periodic patterns of integrated circuit computer memory.

More practical examples of the failure of classical mechanics on an engineering scale are conduction by quantum tunnelling in tunnel diodes and very narrow transistor gates in integrated circuits.

Classical mechanics is the same extreme high frequency approximation as geometric optics. It is more often accurate because it describes particles and bodies with rest mass. These have more momentum and therefore shorter De Broglie wavelengths than massless particles, such as light, with the same kinetic energies.

Branches

Classical mechanics was traditionally divided into three main branches:

- Statics, the study of equilibrium and its relation to forces
- Dynamics, the study of motion and its relation to forces
- Kinematics, dealing with the implications of observed motions without regard for circumstances causing them

Another division is based on the choice of mathematical formalism:

- Newtonian mechanics
- Lagrangian mechanics
- Hamiltonian mechanics

Alternatively, a division can be made by region of application:

- Celestial mechanics, relating to stars, planets and other celestial bodies
- Continuum mechanics, for materials which are modelled as a continuum, e.g., solids and fluids (i.e., liquids and gases).
- Relativistic mechanics (i.e. including the special and general theories of relativity), for bodies whose speed is close to the speed of light.
- Statistical mechanics, which provides a framework for relating the microscopic properties of individual atoms and molecules to the macroscopic or bulk thermodynamic properties of materials.

Newton's laws of motion are three physical laws that form the basis for classical mechanics. They describe the relationship between the forces acting on a body and its motion due to those forces. They have been expressed in several different ways over nearly three centuries, and can be summarised as follows:

1. First law: The velocity of a body remains constant unless the body is acted upon by an external force.
2. Second law: The acceleration a of a body is parallel and directly proportional to the net force F and inversely proportional to the mass m, i.e., $F = ma$.
3. Third law: The mutual forces of action and reaction between two bodies are equal, opposite and collinear.

The three laws of motion were first compiled by Sir Isaac Newton in his work *Philosophiæ Naturalis Principia Mathematica*, first published on July 5, 1687. Newton used them to explain and investigate the motion of many physical objects and systems. For example, in the third volume of the text, Newton showed that these laws of motion, combined with his law of universal gravitation, explained Kepler's laws of planetary motion.

Overview

Newton's laws are applied to bodies (objects) which are considered or idealised as a particle, in the sense that the extent of the body is neglected in the evaluation of its motion, i.e., the object is small compared to the distances involved in the analysis, or the deformation and rotation of the body is of no importance in the analysis. Therefore, a planet can be idealised as a particle for analysis of its orbital motion around a star.

In their original form, Newton's laws of motion are not adequate to characterise the motion of rigid bodies and deformable bodies. Leonard Euler in 1750 introduced a generalisation of Newton's laws of motion for rigid bodies called the Euler's laws of motion, later applied as well for deformable bodies assumed as a continuum. If a body is represented as an assemblage of discrete particles, each governed by Newton's laws of motion, then Euler's laws can be derived from Newton's laws. Euler's laws can, however, be taken as axioms describing the laws of motion for extended bodies, independently of any particle structure.

Newton's Laws hold only with respect to a certain set of frames of reference called Newtonian or inertial reference frames. Some authors interpret the first law as defining what an inertial reference frame is; from this point of view, the second law only holds when the observation is made from an inertial reference frame, and therefore the first law cannot be proved as a special case of the second. Other authors do treat the first law as a corollary of the second. The explicit concept of an inertial frame of reference was not developed until long after Newton's death.

In the given interpretation mass, acceleration, momentum, and (most importantly) force are assumed to be externally defined quantities. This is the most common, but not the only interpretation: one can consider the laws to be a definition of these quantities.

At speeds approaching the speed of light the effects of special relativity must be taken into account.

Newton's First Law

Lex I: Corpus omne perseverare in statu suo quiescendi vel movendi uniformiter in directum, nisi quatenus a viribus impressis cogitur statum illum mutare.

Law I: Every body persists in its state of being at rest or of moving uniformly straight forward, except insofar as it is compelled to change its state by force impressed.

This law states that if the resultant force (the vector sum of all forces acting on an object) is zero, then the velocity of the object is constant. Mathematically stated:

$$\sum F = 0 \Rightarrow \frac{dv}{dt} = 0.$$

Consequently:

- An object that is at rest will stay at rest unless an unbalanced force acts upon it.
- An object that is in motion will not change its velocity unless an unbalanced force acts upon it.

Newton placed the first law of motion to establish frames of reference for which the other laws are applicable. The first law of motion postulates the existence of at least one frame of reference called a Newtonian or inertial reference frame, relative to which the motion of a particle not subject to forces is a straight line at a constant speed. Newton's first law is often referred to as the *law of inertia.* Thus, a condition necessary for the uniform motion of a particle relative to an inertial reference frame is that the total net force acting on it is zero. In this sense, the first law can be restated as:

In every material universe, the motion of a particle in a preferential reference frame Φ is determined by the action of forces whose total vanished for all times when and only when the velocity of the particle is constant in Φ. That is, a particle initially at rest or in uniform motion in the preferential frame Φ continues in that state unless compelled by forces to change it.

Newton's laws are valid only in an inertial reference frame. Any reference frame that is in uniform motion with respect to an inertial frame is also an inertial frame, i.e. Galilean invariance or the principle of Newtonian relativity.

Newton's first law is a restatement of the law of inertia which Galileo had already described and Newton gave credit to Galileo. Aristotle had the view that all objects have a natural place in the universe: that heavy objects like rocks wanted to be at rest on the Earth and that light objects like smoke wanted to be at rest in the sky and the stars wanted to remain in the heavens. He thought that a body was in its natural state when it was at rest, and for the body to move in a straight line at a constant speed an external agent was needed to continually propel it, otherwise it would stop moving. Galileo, however, realised that a force is necessary to change the velocity of a body, i.e., acceleration, but no force is needed to maintain its velocity. This insight leads to Newton's First Law —no force means no acceleration, and hence the body will maintain its velocity.

The law of inertia apparently occurred to several different natural philosophers and scientists independently. The 17th century philosopher René Descartes also formulated the law, although he did not perform any experiments to confirm it.

Newton's Second Law

The second law states that the net force on a particle is equal to the time rate of change of its linear momentum p in aninertial reference frame:

$$F = \frac{dp}{dt} = \frac{d(mv)}{dt},$$

where, since the law is valid only for constant-mass systems, the mass can be taken outside the differentiation operator by the constant factor rule in differentiation. Thus,

$$F = m\frac{dv}{dt} = ma,$$

where F is the net force applied, *m* is the mass of the body, and a is the body's acceleration. Thus, the net force applied to a body produces a proportional acceleration. In other words, if a body is accelerating, then there is a force on it.

Any mass that is gained or lost by the system will cause a change in momentum that is not the result of an external force. A different equation is necessary for variable-mass systems.

Consistent with the first law, the time derivative of the momentum is non-zero when the momentum changes direction, even if there is no change in its magnitude; such is the case with uniform circular motion. The relationship also implies the conservation of momentum: when the net force on the body is zero, the momentum of the body is constant. Any net force is equal to the rate of change of the momentum.

Newton's second law requires modification if the effects of special relativity are to be taken into account, because at high speeds the approximation that momentum is the product of rest mass and velocity is not accurate.

Impulse

An impulse J occurs when a force F acts over an interval of time Δt, and it is given by

$$J = \int_{\Delta t} F dt.$$

Since force is the time derivative of momentum, it follows that

$$J = \Delta p = m\Delta v.$$

This relation between impulse and momentum is closer to Newton's wording of the second law.

Impulse is a concept frequently used in the analysis of collisions and impacts.

Variable-mass Systems

Variable-mass systems, like a rocket burning fuel and ejecting spent gases, are not closed and cannot be directly treated by making mass a function of time in the second law. The reasoning, given in *An Introduction to Mechanics* by Kleppner and Kolenkow and other modern texts, is that Newton's second law applies fundamentally to particles. In classical mechanics, particles by definition have constant mass. In case of a well-defined system of particles, Newton's law can be extended by summing over all the particles in the system:

$$F_{net} = Ma_{cm}$$

where F_{net} is the total external force on the system, M is the total mass of the system, and a_{cm} is the acceleration of the centre of mass of the system.

Variable-mass systems like a rocket or a leaking bucket cannot usually be treated as a system of particles, and thus Newton's second law cannot be applied directly. Instead, the general equation of motion for a body whose mass m varies with time by either ejecting or accreting mass is obtained by rearranging the second law and adding a term to account for the momentum carried by mass entering or leaving the system:

$$F + u\frac{dm}{dt} = m\frac{dv}{dt}$$

where u is the relative velocity of the escaping or incoming mass with respect to the centre of mass of the body.

Under some conventions, the quantity (u dm/dt) on the left-hand side, known as the thrust, is defined as a force (the force exerted on the body by the changing mass, such as rocket exhaust) and is included in the quantity F. Then, by substituting the definition of acceleration, the equation becomes

$$F = ma.$$

History

Newton's original Latin reads:

Lex II: Mutationem motus proportionalem esse vi motrici impressae, et fieri secundum lineam rectam qua vis illa imprimitur.

This was translated quite closely in Motte's 1729 translation as:

Law II: The alteration of motion is ever proportional to the motive force impressed; and is made in the direction of the right line in which that force is impressed.

According to modern ideas of how Newton was using his terminology, this is understood, in modern terms, as an equivalent of:

The change of momentum of a body is proportional to the impulse impressed on the body, and happens along the straight line on which that impulse is impressed.

Motte's 1729 translation of Newton's Latin continued with Newton's commentary on the second law of motion, reading:

If a force generates a motion, a double force will generate double the motion, a triple force triple the motion, whether that force be impressed altogether and at once, or gradually and successively.

And this motion (being always directed the same way with the generating force), if the body moved before, is added to or subtracted from the former motion, according as they directly conspire with or are directly contrary to each other; or obliquely joined, when they are oblique, so as to produce a new motion compounded from the determination of both.

The sense or senses in which Newton used his terminology, and how he understood the second law and intended it to be understood, have been extensively discussed by historians of science, along with the relations between Newton's formulation and modern formulations.

Newton's Third Law

Figure 4: Newton's third law. The skaters' forces on each other are equal in magnitude, but act in opposite directions.

A more direct translation than the one just given above is:

LAW III: To every action there is always opposed an equal reaction: or the mutual actions of two bodies upon each other are always equal, and directed to contrary parts. — Whatever draws or presses another is as much drawn or pressed by that other. If you press a stone with your finger, the finger is also pressed by the stone. If a horse draws a stone tied to a rope, the horse (if I may so say) will be equally drawn back towards the stone: for the distended rope, by the same endeavour to relax or unbend itself, will draw the horse as much towards the stone, as it does the stone towards the horse, and will obstruct the progress of the one as much as it advances that of the other. If a body impinges upon another, and by its force changes the motion of the other, that body also (because of the equality of the mutual pressure) will undergo an equal change, in its own motion, toward the contrary part. The changes made by these actions are equal, not in the velocities but in the motions of the bodies; that is to say, if the bodies are not hindered by any other impediments. For, as the motions are equally changed, the changes of the velocities made toward contrary parts are reciprocally proportional to the bodies. This law takes place also in attractions, as will be proved in the next scholium.

In the above, as usual, *motion* is Newton's name for momentum, hence his careful distinction between motion and velocity.

The Third Law means that all forces are *interactions* between different bodies, and thus that there is no such thing as a unidirectional force or a force that acts on only one body. Whenever a first body exerts a force F on a second body, the second body exerts a force -F on the first body. F and -F are equal in magnitude and opposite in direction. This law is sometimes referred to as the *action-reaction law*, with F called the "action" and -F the "reaction". The action and the reaction are simultaneous.

As shown in the diagram opposite, the skaters' forces on each other are equal in magnitude, but act in opposite directions. Although the forces are equal, the accelerations are not: the less massive skater will have a greater acceleration due to Newton's second law. The two forces in Newton's third law are of the same type (e.g., if the road exerts a forward frictional force on an accelerating car's tires, then it is also a frictional force that Newton's third law predicts for the tires pushing backward on the road). Put very simply: a force acts between a pair of objects, and not on a single object. So each and every force has two ends. Each of the two ends is the same except for being opposite in direction. The ends of a force are mirror images of each other, one might say. From a mathematical point of view, Newton's third law is a one-dimensional vector equation, which can be stated as follows. Given two objects A and B, each exerting a force on the other,

$$\sum F_{a,b} = -\sum F_{b,a}$$

where

$F_{a,b}$ are the forces from B acting on A, and

$F_{b,a}$ are the forces from A acting on B.

Newton used the third law to derive the law of conservation of momentum; however from a deeper perspective, conservation of momentum is the more fundamental idea (derived via Noether's theorem from Galilean invariance), and holds in cases where Newton's third law appears to fail, for instance when force fields as well as particles carry momentum, and in quantum mechanics.

Importance and Range of Validity

Newton's laws were verified by experiment and observation for over 200 years, and they are excellent approximations at the scales and speeds of everyday life. Newton's laws of motion, together with his law of universal gravitation and the mathematical techniques of calculus, provided for the first time a unified quantitative explanation for a wide range of physical phenomena.

These three laws hold to a good approximation for macroscopic objects under everyday conditions. However, Newton's laws (combined with universal gravitation and classical electrodynamics) are inappropriate for use in certain circumstances, most notably at very small scales, very high speeds (in special relativity, the Lorentz factor must be included in the expression for momentum along with rest mass and velocity) or very strong gravitational fields. Therefore, the laws cannot be used to explain phenomena such as conduction of electricity in a semiconductor, optical properties of substances, errors in non-relativistically corrected GPS systems and superconductivity. Explanation of these phenomena requires more sophisticated physical theories, including general relativity and quantum field theory.

In quantum mechanics concepts such as force, momentum, and position are defined by linear operators that operate on the quantum state; at speeds that are much lower than the speed of light, Newton's laws are just as exact for these operators as they are for classical objects. At speeds comparable to the speed of light, the second law holds in the original form $\mathrm{F} = \mathrm{dp} \;\; \mathrm{d}t$, which says that the force is the derivative of the momentum of the object with respect to time, but some of the newer versions of the second law (such as the constant mass approximation above) do not hold at relativistic velocities.

Relationship to the Conservation Laws

In modern physics, the laws of conservation of momentum, energy, and angular momentum are of more general validity than Newton's laws, since they apply to both light and matter, and to both classical and non-classical physics.

This can be stated simply, "Momentum, energy and angular momentum cannot be created or destroyed."

Because force is the time derivative of momentum, the concept of force is redundant and subordinate to the conservation of momentum, and is not used in fundamental theories (e.g., quantum mechanics, quantum electrodynamics, general relativity, etc.). The standard model explains in detail how the three fundamental forces known as gauge forces originate out of exchange by virtual particles. Other forces such as gravity and fermionic degeneracy pressure also arise from the momentum conservation. Indeed, the conservation of 4-momentum in inertial motion via curved space-time results in what we call gravitational force in general relativity theory. Application of space derivative (which is a momentum operator in quantum mechanics) to overlapping wave functions of pair of fermions (particles with semi-integer spin) results in shifts of maxima of compound wave function away from each other, which is observable as "repulsion" of fermions.

Newton stated the third law within a world-view that assumed instantaneous action at a distance between material particles. However, he was prepared for philosophical criticism of this action at a distance, and it was in this context that he stated the famous phrase "I feign no hypotheses". In modern physics, action at a distance has been completely eliminated, except for subtle effects involving quantum entanglement. However in modern engineering in all practical applications involving the motion of vehicles and satellites, the concept of action at a distance is used extensively. Conservation of energy was discovered nearly two centuries after Newton's lifetime, the long delay occurring because of the difficulty in understanding the role of microscopic and invisible forms of energy such as heat and infrared light.

Lagrangian Mechanics

Lagrangian mechanics is a re-formulation of classical mechanics that combines conservation of momentum with conservation of energy. It was introduced by the Italian-French mathematician Joseph-Louis Lagrange in 1788. In Lagrangian mechanics, the trajectory of a system of particles is derived by solving the Lagrange equations in one of two forms, either the Lagrange equations of the first kind, which treat constraints explicitly as extra equations, often using Lagrange multipliers; or the Lagrange equations of the second kind, which incorporate the constraints directly by judicious choice of generalised coordinates. The fundamental lemma of the calculus of variations shows that solving the Lagrange equations is equivalent to finding the path for which the action functional is stationary, a quantity that is the integral of the Lagrangian over time.

The use of generalised coordinates may considerably simplify a system's analysis. For example, consider a small frictionless bead travelling in a groove. If one is tracking the bead as a particle, calculation of the motion of the bead using Newtonian mechanics would require solving for the time-varying constraint force required to keep the bead in the groove. For the same problem using Lagrangian mechanics, one looks at the path of the groove and chooses a set of *independent generalised* coordinates that completely characterise the possible motion of the bead. This choice eliminates the need for the constraint force to enter into the resultant system of equations. There are fewer equations since one is not directly calculating the influence of the groove on the bead at a given moment. [This does not appear to be true, since using standard Newtonian mechanics one would ignore an analysis of forces and use conservation of energy which follows directly from Newton's laws; alternatively one can eliminate the constraint force by resolving tangentially to the groove].

Lagrange Equations of the Second Kind

The equations of motion in Lagrangian mechanics are the *Lagrange equations*, also known as the Euler–*Lagrange equations*. In lagrange equation there are 2n Generalised coordinates. Below, we sketch out the derivation of the Lagrange equations of the second kind. Please note that in this context, V is used rather than U for potential energy and T replaces K for kinetic energy.

Start with D'Alembert's principle for the virtual work of applied forces, F_i, and inertial forces on a three dimensional accelerating system of n particles, i, whose motion is consistent with its constraints,

$$\delta W = \sum_{i=1}^{n} (F_i - m_i a_i) \cdot \delta r_i = 0.$$

where

δW is the virtual work;

δr_i is the virtual displacement of the system, consistent with the constraints;

m_i are the masses of the particles in the system;

a_i are the accelerations of the particles in the system;

$m_i a_i$ together as products represent the time derivatives of the system momenta, aka. inertial forces;

i is an integer used to indicate (via subscript) a variable corresponding to a particular particle; and

n is the number of particles under consideration.

Break out the two terms:

$$\delta W = \sum_{i=1}^{n} F_i \cdot \delta r_i - \sum_{i=1}^{n} m_i a_i \cdot \delta r_i = 0.$$

Assume that the following transformation equations from m independent generalised coordinates, q_j, hold:

$$\begin{aligned} r_1 &= r_1(q_1, q_2, \ldots, q_m, t) \\ r_2 &= r_2(q_1, q_2, \ldots, q_m, t) \\ &\vdots \\ r_n &= r_n(q_1, q_2, \ldots, q_m, t) \end{aligned}$$

where m (without a subscript) indicates the total number of generalised coordinates. An expression for the virtual displacement (differential), δr_i of the system for *time-independent constraints* or "velocity-dependent constraints" is

$$\delta r_i = \sum_{j=1}^{m} \frac{\partial r_i}{\partial q_j} \delta q_j,$$

where j is an integer used to indicate (via subscript) a variable corresponding to a generalised coordinate.

The applied forces may be expressed in the generalised coordinates as generalised forces, Q_j:

$$Q_j = \sum_{i=1}^{n} F_i \cdot \frac{\partial r_i}{\partial q_j}.$$

Combining the equations for δW, δr_i, and Q_j yields the following result after pulling the sum out of the dot product in the second term:

$$\delta W = \sum_{j=1}^{m} Q_j \delta q_j - \sum_{j=1}^{m} \sum_{i=1}^{n} m_i a_i \cdot \frac{\partial r_i}{\partial q_j} \delta q_j = 0.$$

Substituting in the result from the kinetic energy relations to change the inertial forces into a function of the kinetic energy leaves

$$\delta W = \sum_{j=1}^{m} Q_j \delta q_j - \sum_{j=1}^{m} \left(\frac{d}{dt}\left(\frac{\partial T}{\partial \dot{q}_j} \right) - \frac{\partial T}{\partial q_j} \right) \delta q_j = 0.$$

In the above equation, δq_j is arbitrary, though it is by definition consistent with the constraints. So the relation must hold term-wise:

$$Q_j = \frac{d}{dt}\left(\frac{\partial T}{\partial \dot{q}_j} \right) - \frac{\partial T}{\partial q_j}.$$

If the F_i are conservative, they may be represented by a scalar potential field, V:

$$F_i = -\nabla V \Rightarrow Q_j = -\sum_{i=1}^{n} \nabla V \cdot \frac{\partial r_i}{\partial q_j} = -\frac{\partial V}{\partial q_j}.$$

The previous result may be easier to see by recognising that V is a function of the r_i, which are in turn functions of q_j, and then applying the chain rule to the derivative of V with respect to q_j.

Recall the definition of the Lagrangian is

$$L = T - V.$$

Since the potential field is only a function of position, not velocity, Lagrange's equations are as follows:

$$\frac{d}{dt}\left(\frac{\partial L}{\partial \dot{q}_j}\right) - \frac{\partial L}{\partial q_j} = \frac{d}{dt}\left(\frac{\partial T}{\partial \dot{q}_j}\right) - \frac{\partial T}{\partial q_j} + \frac{\partial V}{\partial q_j} = 0.$$

This is consistent with the results derived above and may be seen by differentiating the right side of the Lagrangian with respect to $\dot{q}_j$ and time, and solely with respect to q_j, adding the results and associating terms with the equations for F and Q_j.

In a more general formulation, the forces could be both potential and viscous. If an appropriate transformation can be found from the F_i, Rayleigh suggests using a dissipation function, D, of the following form:

$$D = \frac{1}{2}\sum_{j=1}^{m}\sum_{k=1}^{m} C_{jk}\dot{q}_j\dot{q}_k.$$

where C_{jk} are constants that are related to the damping coefficients in the physical system, though not necessarily equal to them

If D is defined this way, then

$$Q_j = -\frac{\partial V}{\partial q_j} - \frac{\partial D}{\partial \dot{q}_j}$$

and

$$0 = \frac{d}{dt}\left(\frac{\partial L}{\partial \dot{q}_j}\right) - \frac{\partial L}{\partial q_j} + \frac{\partial D}{\partial \dot{q}_j}.$$

Kinetic Energy Relations

The kinetic energy, T, for the system of particles is defined by

$$T = \frac{1}{2}\sum_{i=1}^{n} m_i v_i \cdot v_i.$$

The partial derivative of T with respect to the time derivatives of the generalised coordinates, $\dot{q}_j$, is

$$\frac{\partial T}{\partial \dot{q}_j} = \sum_{i=1}^{n} m_i v_i \cdot \frac{\partial v_i}{\partial \dot{q}_j}.$$

The previous result may be difficult to visualise. As a result of the product rule, the derivative of a general dot product is

$$\frac{d}{dx}(f(x)\cdot g(x)) = f(x)\cdot\frac{d}{dx}g(x) + g(x)\cdot\frac{d}{dx}f(x).$$

This general result may be seen by briefly stepping into a Cartesian coordinate system, recognising that the dot product is (there) a term-by-term product sum, and also recognising that the derivative of a sum is the sum of its derivatives. In our case, f and g are equal to v, which is why the factor of one half disappears.

According to the chain rule and the coordinate transformation equations given above for r, its time derivative, v, is

$$v_i = \sum_{k=1}^{m}\frac{\partial r_i}{\partial q_k}\dot{q}_k + \frac{\partial r_i}{\partial t}.$$

Together, the definition of v_i and the total differential, dr_i, suggest that

$$\frac{\partial v_i}{\partial \dot{q}_j} = \frac{\partial r_i}{\partial q_j}$$

since

$$\frac{\partial T}{\partial \dot{q}_j} = \sum_{i=1}^{n} m_i v_i \cdot \frac{\partial r_i}{\partial q_j}. \quad \frac{\partial}{\partial \dot{q}_k}A\dot{q}_k = A$$

and that in the sum, there is only one $\dot{q}_j$.

Substituting this relation back into the expression for the partial derivative of *T* gives

Taking the time derivative gives

$$\frac{d}{dt}\left(\frac{\partial T}{\partial \dot{q}_j}\right) = \sum_{i=1}^{n}\left[m_i a_i \cdot \frac{\partial r_i}{\partial q_j} + m_i v_i \cdot \frac{d}{dt}\left(\frac{\partial r_i}{\partial q_j}\right)\right].$$

Using the chain rule on the last term gives

$$\frac{d}{dt}\left(\frac{\partial r_i}{\partial q_j}\right) = \sum_{k=1}^{m}\frac{\partial^2 r_i}{\partial q_j \partial q_k}\dot{q}_k + \frac{\partial^2 r_i}{\partial q_j \partial t}.$$

From the expression for v_i, one sees that also

$$\frac{\partial v_i}{\partial q_j} = \sum_{k=1}^{m}\frac{\partial^2 r_i}{\partial q_j \partial q_k}\dot{q}_k + \frac{\partial^2 r_i}{\partial q_j \partial t}.$$

This allows simplification of the last term,

$$\frac{d}{dt}\left(\frac{\partial T}{\partial \dot{q}_j}\right)=\sum_{i=1}^{n}\left[m_i a_i\cdot\frac{\partial r_i}{\partial q_j}+m_i v_i\cdot\frac{\partial v_i}{\partial q_j}\right].$$

The partial derivative of T with respect to the generalised coordinates, q_j, is

$$\frac{\partial T}{\partial q_j}=\sum_{i=1}^{n}\frac{\partial[\frac{1}{2}m_i v_i^2]}{\partial q_j}=\sum_{i=1}^{n}\frac{\partial[\frac{1}{2}m_i\,(v_i\cdot v_i)]}{\partial q_j}=\frac{1}{2}\sum_{i=1}^{n}\left[m_i v_i\cdot\frac{\partial v_i}{\partial q_j}+m_i v_i\cdot\frac{\partial v_i}{\partial q_j}\right]$$

This last result may be obtained by doing a partial differentiation directly on the kinetic energy definition represented by the first equation. The last two equations may be combined to give an expression for the inertial forces in terms of the kinetic energy:

$$\frac{d}{dt}\left(\frac{\partial T}{\partial \dot{q}_j}\right)-\frac{\partial T}{\partial q_j}=\sum_{i=1}^{n}m_i a_i\cdot\frac{\partial r_i}{\partial q_j}.$$

Old Lagrange's Equations

Consider a single particle with mass m and position vector r, moving under an applied force, F, which can be expressed as the gradient of a scalar potential energy function V(r,t):

$$F=-\nabla V.$$

Such a force is independent of third- or higher-order derivatives of r, so Newton's second law forms a set of 3 second-order ordinary differential equations. Therefore, the motion of the particle can be completely described by 6 independent variables, or *degrees of freedom*.

An obvious set of variables is $\{r_j,\dot{r}_j \mid j=1,2,3\}$, the Cartesian components of r and their time derivatives, at a given instant of time (i.e. position (x,y,z) and velocity (v_x,v_y,v_z)).

More generally, we can work with a set of generalised coordinates, q_j, and their time derivatives, the generalised velocities, $\dot{q}_j$. The position vector, r, is related to the generalised coordinates by some transformation *equation*:

$$r=r(q_i,q_j,q_k,t).$$

For example, for a simple pendulum of length ℓ, a logical choice for a generalised coordinate is the angle of the pendulum from vertical, θ, for which the transformation equation would be

$$r(\theta,\dot{\theta},t) = (\ell \sin\theta, \ell \cos\theta).$$

The term "generalised coordinates" is really a holdover from the period when Cartesian coordinates were the default coordinate system. Consider an arbitrary displacement δr of the particle. The work done by the applied force F is $W = F\cdot\delta r$. Using Newton's second law, we write:

$$F\cdot\delta r = m\ddot{r}\cdot\delta r.$$

Since work is a physical scalar quantity, we should be able to rewrite this equation in terms of the generalised coordinates and velocities. On the left hand side,

$$F\cdot\delta r \quad = -\nabla V\cdot\sum_i \frac{\partial r}{\partial q_i}\delta q_i$$

$$[8pt] \quad = -\sum_{i,j}\frac{\partial V}{\partial r_j}\frac{\partial r_j}{\partial q_i}\delta q_i$$

$$[8pt] \quad = -\sum_i \frac{\partial V}{\partial q_i}\delta q_i.$$

On the right hand side, carrying out a change of coordinates to generalised coordinates, we obtain:

$$m\ddot{r}\cdot\delta r = m\sum_{i,j}\ddot{r}_i\frac{\partial r_i}{\partial q_j}\delta q_j$$

Rearranging slightly:

$$m\ddot{r}\cdot\delta r = m\sum_j\left[\sum_i \ddot{r}_i\frac{\partial r_i}{\partial q_j}\right]\delta q_j$$

Now, by performing an "integration by parts" transformation, with respect to t:

$$m\ddot{r}\cdot\delta r = m\sum_j\left[\sum_i\left[\frac{d}{dt}\left(\dot{r}_i\frac{\partial r_i}{\partial q_j}\right)-\dot{r}_i\frac{d}{dt}\left(\frac{\partial r_i}{\partial q_j}\right)\right]\right]\delta q_j$$

Recognising that $\frac{d}{dt}\frac{\partial r_j}{\partial q_i} = \frac{\partial \dot{r}_j}{\partial q_i}$ and $\frac{\partial r_j}{\partial q_i} = \frac{\partial \dot{r}_j}{\partial \dot{q}_i}$, we obtain:

$$m\ddot{r}\cdot\delta r = m\sum_j\left[\sum_i\left[\frac{d}{dt}\left(\dot{r}_i\frac{\partial \dot{r}_i}{\partial \dot{q}_j}\right)-\dot{r}_i\frac{\partial \dot{r}_i}{\partial q_j}\right]\right]\delta q_j$$

Now, by changing the order of differentiation, we obtain:

$$m\ddot{r}\cdot\delta r = m\sum_j\left[\sum_i\left[\frac{d}{dt}\frac{\partial}{\partial \dot{q}_j}\left(\frac{1}{2}\dot{r}_i^{2}\right)-\frac{\partial}{\partial q_j}\left(\frac{1}{2}\dot{r}_i^{2}\right)\right]\right]\delta q_j$$

Finally, we change the order of summation:

$$m\ddot{r}\cdot\delta r = \sum_j\left[\frac{d}{dt}\frac{\partial}{\partial \dot{q}_j}\left(\sum_i\frac{1}{2}m\dot{r}_i^{2}\right)-\frac{\partial}{\partial q_j}\left(\sum_i\frac{1}{2}m\dot{r}_i^{2}\right)\right]\delta q_j$$

Which is equivalent to:

$$m\ddot{r}\cdot\delta r = \sum_i\left[\frac{d}{dt}\frac{\partial T}{\partial \dot{q}_i}-\frac{\partial T}{\partial q_i}\right]\delta q_i$$

where $T=\frac{1}{2}m\dot{r}\cdot\dot{r}$ is the kinetic energy of the particle. Our equation for the work done becomes

$$\sum_i\left[\frac{d}{dt}\frac{\partial T}{\partial \dot{q}_i}-\frac{\partial (T-V)}{\partial q_i}\right]\delta q_i = 0.$$

However, this must be true for *any* set of generalised displacements δq_i, so we must have

$$\left[\frac{d}{dt}\frac{\partial T}{\partial \dot{q}_i}-\frac{\partial (T-V)}{\partial q_i}\right] = 0$$

for *each* generalised coordinate δq_i. We can further simplify this by noting that *V* is a function solely ofr and *t*, and r is a function of the generalised coordinates and *t*. Therefore, *V* is independent of the generalised velocities:

$$\frac{d}{dt}\frac{\partial V}{\partial \dot{q}_i} = 0.$$

Inserting this into the preceding equation and substituting *L=T-V*, called the Lagrangian, we obtain Lagrange's equations:

$$\frac{\partial L}{\partial q_i} = \frac{d}{dt}\frac{\partial L}{\partial \dot{q}_i}.$$

There is one Lagrange equation for each generalised coordinate q_i. When $q_i = r_i$ (i.e. the generalised coordinates are simply the Cartesian coordinates), it is straightforward to check that Lagrange's equations reduce to Newton's second law.

The above derivation can be generalised to a system of N particles. There will be $6N$ generalised coordinates, related to the position coordinates by $3N$ transformation equations. In each of the $3N$ Lagrange equations, T is the total kinetic energy of the system, and V the total potential energy. In practice, it is often easier to solve a problem using the Euler–Lagrange equations than Newton's laws. This is because not only may more appropriate generalised coordinates q_i be chosen to exploit symmetries in the system, but constraint forces are replaced with simpler relations.

Examples

In this section two examples are provided in which the above concepts are applied. The first example establishes that in a simple case, the Newtonian approach and the Lagrangian formalism agree. The second case illustrates the power of the above formalism, in a case which is hard to solve with Newton's laws.

Falling Mass

Consider a point mass m falling freely from rest. By gravity a force $F = mg$ is exerted on the mass (assuming g constant during the motion). Filling in the force in Newton's law, we find $\ddot{x} = g$ from which the solution

$$x(t) = \frac{1}{2} g t^2$$

follows (choosing the origin at the starting point). This result can also be derived through the Lagrangian formalism. Take x to be the coordinate, which is *0* at the starting point. The kinetic energy is $T = {}^{1}\mathrm{D}$ ${}_{2}mv^2$ and the potential energy is $V =$ "mgx; hence,

$$L = T - V = \frac{1}{2} m\dot{x}^2 + mgx..$$

Then

$$0 = \frac{\partial L}{\partial x} - \frac{d}{dt}\frac{\partial L}{\partial \dot{x}} = mg - m\frac{d\dot{x}}{dt}$$

which can be rewritten as $\ddot{x} = g$, yielding the same result as earlier.

Pendulum on a Movable Support

Consider a pendulum of mass m and length ℓ, which is attached to a support with mass M which can move along a line in the x-direction. Let x be the coordinate along the line of the support, and let us denote the position of the pendulum by the angle θ from the vertical.

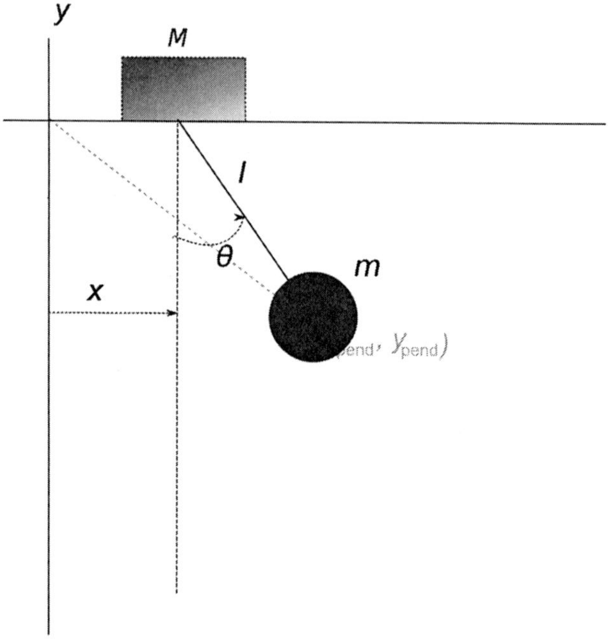

Figure 5: Sketch of the situation with definition of the coordinates

The kinetic energy can then be shown to be

$$\begin{aligned} T &= \frac{1}{2}M\dot{x}^2 + \frac{1}{2}m\left(\dot{x}_{pend}^2 + \dot{y}_{pend}^2\right) \\ &= \frac{1}{2}M\dot{x}^2 + \frac{1}{2}m\left[\left(\dot{x} + \ell\dot{\theta}\cos\theta\right)^2 + \left(\ell\dot{\theta}\sin\theta\right)^2\right], \end{aligned}$$

and the potential energy of the system is

$$V = mg\,y_{pend} = -mg\ell\cos\theta.$$

The Lagrangian is therefore

$$\begin{aligned} L &= T - V \\ &= \frac{1}{2}M\dot{x}^2 + \frac{1}{2}m\left[\left(\dot{x} + \ell\dot{\theta}\cos\theta\right)^2 + \left(\ell\dot{\theta}\sin\theta\right)^2\right] + mg\ell\cos\theta \\ &= \frac{1}{2}(M+m)\dot{x}^2 + m\dot{x}\ell\dot{\theta}\cos\theta + \frac{1}{2}m\ell^2\dot{\theta}^2 + mg\ell\cos\theta \end{aligned}$$

Now carrying out the differentiations gives for the support coordinate x

$$\frac{d}{dt}\left[(M+m)\dot{x}+m\ell\dot{\theta}\cos\theta\right]=0,$$

therefore:

$$(M+m)\ddot{x}+m\ell\ddot{\theta}\cos\theta-m\ell\dot{\theta}^2\sin\theta=0$$

indicating the presence of a constant of motion. Performing the same procedure for the variable θ yields:

$$\frac{d}{dt}\left[m(\dot{x}\ell\cos\theta+\ell^2\dot{\theta})\right]+m(\dot{x}\ell\dot{\theta}+g\ell)\sin\theta=0;$$

therefore

$$\ddot{\theta}+\frac{\ddot{x}}{\ell}\cos\theta+\frac{g}{\ell}\sin\theta=0.$$

These equations may look quite complicated, but finding them with Newton's laws would have required carefully identifying all forces, which would have been much harder and more prone to errors. By considering limit cases (for example, $\ddot{x}\to 0$ should give the equations of motion for a pendulum which is at rest in some inertial frame, while $\ddot{\theta}\to 0$ should give the equations for a pendulum in a constantly accelerating system, etc.) the correctness of this system can be verified. Furthermore, it is trivial to obtain the results numerically, given suitable starting conditions and a chosen time step, by stepping through the results iteratively.

Two-body Central Force Problem

The basic problem is that of two bodies in orbit about each other attracted by a central force. The Jacobi coordinates are introduced; namely, the location of the centre of mass R and the separation of the bodies r (the relative position). The Lagrangian is then

$$\begin{aligned} L &= T-U=\frac{1}{2}M\dot{R}^2+\left(\frac{1}{2}\mu\dot{r}^2-U(r)\right) \\ &= L_{cm}+L_{rel} \end{aligned}$$

where M is the total mass, μ is the reduced mass, and U the potential of the radial force. The Lagrangian is divided into a *centre-of-mass* term and a *relative motion* term. The R equation from the Euler-Lagrange system is simply:

$$M\ddot{R}=0,$$

resulting in simple motion of the centre of mass in a straight line at constant velocity. The relative motion is expressed in polar coordinates (r, θ):

$$L = \frac{1}{2}\mu\left(\dot{r}^2 + r^2\dot{\theta}^2\right) - U(r),$$

which does not depend upon θ, therefore an *ignorable* coordinate. The Lagrange equation for θ is then:

$$\frac{\partial L}{\partial \dot{\theta}} = \mu r^2 \dot{\theta} = \text{constant} = \ell,$$

where ℓ is the conserved angular momentum. The Lagrange equation for r is:

$$\frac{\partial L}{\partial r} = \frac{d}{dt}\frac{\partial L}{\partial \dot{r}},$$

or:

$$\mu r\dot{\theta}^2 - \frac{dU}{dr} = \mu \ddot{r}.$$

This equation is identical to the radial equation obtained using Newton's laws in a *co-rotating* reference frame, that is, a frame rotating with the reduced mass so it appears stationary. If the angular velocity is replaced by its value in terms of the angular momentum,

$$\mu r\dot{\theta}^2 - \frac{dU}{dr} = \mu \ddot{r}.$$

the radial equation becomes:

$$\mu \ddot{r} = -\frac{dU}{dr} + \frac{\ell^2}{\mu r^3}.$$

which is the equation of motion for a one-dimensional problem in which a particle of mass μ is subjected to the inward central force "dU/dr and a second outward force, called in this context the centrifugal force:

$$F_{cf} = \mu r\dot{\theta}^2 = \frac{\ell^2}{\mu r^3}.$$

Of course, if one remains entirely within the one-dimensional formulation, ℓ enters only as some imposed parameter of the external outward force, and its interpretation as angular momentum depends upon the more general two-dimensional problem from which the one-dimensional problem originated.

If one arrives at this equation using Newtonian mechanics in a co-rotating frame, the interpretation is evident as the centrifugal force in that frame due to the rotation of the frame itself. If one arrives at this equation directly by using the generalised coordinates (r, θ) and simply following the Lagrangian formulation without thinking about frames at all, the interpretation is that the centrifugal force is an outgrowth of *using polar coordinates*. As Hildebrand says: "Since such quantities are not true physical forces, they are often called *inertia forces*. Their presence or absence depends, not upon the particular problem at hand, but *upon the coordinate system chosen*." In particular, if Cartesian coordinates are chosen, the centrifugal force disappears, and the formulation involves only the central force itself, which provides the centripetal force for a curved motion.

This viewpoint, that fictitious forces originate in the choice of coordinates, often is expressed by users of the Lagrangian method. This view arises naturally in the Lagrangian approach, because the frame of reference is (possibly unconsciously) selected by the choice of coordinates. Unfortunately, this usage of "inertial force" conflicts with the Newtonian idea of an inertial force. In the Newtonian view, an inertial force originates in the acceleration of the frame of observation (the fact that it is not an inertial frame of reference), not in the choice of coordinate system. To keep matters clear, it is safest to refer to the Lagrangian inertial forces as *generalised* inertial forces, to distinguish them from the Newtonian vector inertial forces. That is, one should avoid following Hildebrand when he says "we deal *always* with *generalised* forces, velocities accelerations, and momenta. For brevity, the adjective "generalised" frequently will be omitted."

It is known that the Lagrangian of a system is not unique. Within the Lagrangian formalism the Newtonian fictitious forces can be identified by the existence of alternative Lagrangians in which the fictitious forces disappear, sometimes found by exploiting the symmetry of the system.

Hamilton's Principle

The action, denoted by S, is the time integral of the Lagrangian:

$$S = \int L\,dt.$$

Let q_0 and q_1 be the coordinates at respective initial and final times t_0 and t_1. Using the calculus of variations, it can be shown the Lagrange's equations are equivalent to *Hamilton's principle*:

The system undergoes the trajectory between t_0 and t_1 whose action has a stationary value.

By *stationary*, we mean that the action does not vary to first-order for infinitesimal deformations of the trajectory, with the end-points (q_0, t_0) and (q_1, t_1) fixed. Hamilton's principle can be written as:

$$\delta S = 0.$$

Thus, instead of thinking about particles accelerating in response to applied forces, one might think of them picking out the path with a stationary action.

Hamilton's principle is sometimes referred to as the *principle of least action*. However, this is a misnomer: the action only needs to be stationary, with any variation h of the functional giving an increase in the functional integral of the action. This is not, as is frequently misstated, required to be a maximum or a minimum of the action functional.

We can use this principle instead of Newton's Laws as the fundamental principle of mechanics, this allows us to use an integral principle (Newton's Laws are based on differential equations so they are a differential principle) as the basis for mechanics. However it is not widely stated that Hamilton's principle is a variational principle only with holonomic constraints, if we are dealing with nonholonomic systems then the variational principle should be replaced with one involving d'Alembert principle of virtual work. Working only with holonomic constraints is the price we have to pay for using an elegant variational formulation of mechanics.

Extensions of Lagrangian Mechanics

The Hamiltonian, denoted by H, is obtained by performing a Legendre transformation on the Lagrangian, which introduces new variables, canonically conjugate to the original variables. This doubles the number of variables, but makes differential equations first order. The Hamiltonian is the basis for an alternative formulation of classical mechanics known as Hamiltonian mechanics. It is a particularly ubiquitous quantity in quantum mechanics.

In 1948, Feynman discovered the path integral formulation extending the principle of least action toquantum mechanics for electrons and photons. In this formulation, particles travel every possible path between the initial and final states; the probability of a specific final state is obtained by summing over all possible trajectories leading to it. In the classical regime, the path integral formulation cleanly reproduces Hamilton's principle, and Fermat's principle in optics.

Hamiltonian Mechanics

Hamiltonian mechanics is a reformulation of classical mechanics that was introduced in 1833 by Irish mathematician William Rowan Hamilton.

It arose from Lagrangian mechanics, a previous reformulation of classical mechanics introduced by Joseph Louis Lagrange in 1788, but can be formulated without recourse to Lagrangian mechanics using symplectic spaces.

The Hamiltonian method differs from the Lagrangian method in that instead of expressing second-order differential constraints on an n-dimensional coordinate space (where n is the number of degrees of freedom of the system), it expresses first-order constraints on a $2n$-dimensional phase space.

As with Lagrangian mechanics, Hamilton's equations provide a new and equivalent way of looking at classical mechanics. Generally, these equations do not provide a more convenient way of solving a particular problem. Rather, they provide deeper insights into both the general structure of classical mechanics and its connection to quantum mechanics as understood through Hamiltonian mechanics, as well as its connection to other areas of science.

Simplified Overview of Uses

The value of the Hamiltonian is the total energy of the system being described. For a closed system, it is the sum of the kinetic and potential energy in the system. There is a set of differential equations known as the *Hamilton equations* which give the time evolution of the system. Hamiltonians can be used to describe such simple systems as a bouncing ball, a pendulum or an oscillating spring in which energy changes from kinetic to potential and back again over time. Hamiltonians can also be employed to model the energy of other more complex dynamic systems such as planetary orbits incelestial mechanics and also in quantum mechanics.

The Hamilton equations are generally written as follows:

$$\dot{p} = -\frac{\partial H}{\partial q}$$

$$\dot{q} = \frac{\partial H}{\partial p}.$$

In the above equations, the dot denotes the ordinary derivative with respect to time of the functions $p = p(t)$ (called generalised momenta) and $q = q(t)$ (called generalised coordinates), taking values in some vector space, and $H = H(p,q,t)$ is the so-called Hamiltonian, or (scalar valued) Hamiltonian function. Thus, more explicitly, one can equivalently write

$$\frac{d}{dt}p(t) = -\frac{\partial}{\partial q}H(p(t),q(t),t)$$

$$\frac{d}{dt}q(t) = \frac{\partial}{\partial p}H(p(t),q(t),t)$$

and specify the domain of values in which the parameter t (time) varies.

Basic Physical Interpretation

The simplest interpretation of the Hamilton equations is as follows, applying them to a one-dimensional system consisting of one particle of mass m under time-independent boundary conditions: The Hamiltonian H represents the energy of the system (provided that there are NO external forces, or additional energy added to the system), which is the sum of kinetic and potential energy, traditionally denoted T and V, respectively. Here q is the x coordinate and p is the momentum, mv. Then

$$H = T + V, \quad T = \frac{p^2}{2m}, \quad V = V(q) = V(x).$$

Note that T is a function of p alone, while V is a function of x (or q) alone.

Now the time-derivative of the momentum p equals the *Newtonian force*, and so here the first Hamilton equation means that the force on the particle equals the rate at which it loses potential energy with respect to changes in x, its location. (Force equals the negative gradient of potential energy.)

The time-derivative of q here means the velocity: the second Hamilton equation here means that the particle's velocity equals the derivative of its kinetic energy with respect to its momentum. (Because the derivative with respect to p of $p^2/2m$ equals $p/m = mv/m = v$.)

Using Hamilton's Equations

In terms of the generalised coordinates q_i and generalised velocities $\dot{q}_i$:

1. First write out the Lagrangian $L = T - V$. Express *T* and *V* as though you were going to use Lagrange's equation.
2. Calculate the momenta by differentiating the Lagrangian with respect to velocity: $p_i(q_i, \dot{q}_i, t) = \frac{\partial L}{\partial \dot{q}_i}$.
3. Express the velocities in terms of the momenta by inverting the expressions in step (2).
4. Calculate the Hamiltonian using the usual definition of H as the Legendre transformation of L: $H = \sum_i \dot{q}_i \frac{\partial L}{\partial \dot{q}_i} - L = \sum_i \dot{q}_i p_i - L$. Substitute for the velocities using the results in step (3).
5. Apply Hamilton's equations.

Notes

Hamilton's equations are appealing in view of their beautiful simplicity and (slightly *broken*) symmetry. They have been analysed under almost every imaginable perspective, from basic physics up to symplectic geometry. A lot is known about solutions of these equations, yet the exact general case solution of the equations of motion cannot be given explicitly for a system of more than two massive point particles. The finding of conserved quantities plays an important role in the search for solutions or information about their nature. In models with an infinite number of degrees of freedom, this is of course even more complicated. An interesting and promising area of research is the study of integrable systems, where an infinite number of independent conserved quantities can be constructed.

Deriving Hamilton's Equations

We can derive Hamilton's equations by looking at how the total differential of the Lagrangian depends on time, generalised positions q_i and generalised velocities $\dot{q}_i$:

$$dL = \sum_i \left(\frac{\partial L}{\partial q_i} dq_i + \frac{\partial L}{\partial \dot{q}_i} d\dot{q}_i \right) + \frac{\partial L}{\partial t} dt.$$

Now the generalised momenta were defined as $p_i = \frac{\partial L}{\partial \dot{q}_i}$ and Lagrange's equations tell us that

$$\frac{d}{dt} \frac{\partial L}{\partial \dot{q}_i} - \frac{\partial L}{\partial q_i} = 0.$$

We can rearrange this to get

$$\frac{\partial L}{\partial q_i} = \dot{p}_i$$

and substitute the result into the total differential of the Lagrangian

$$dL = \sum_i [\dot{p}_i dq_i + p_i d\dot{q}_i] + \frac{\partial L}{\partial t} dt.$$

We can rewrite this as

$$dL = \sum_i [\dot{p}_i dq_i + p_i d\dot{q}_i] + \frac{\partial L}{\partial t} dt.$$

and rearrange again to get

$$d\left(\sum_i p_i \dot{q}_i - L\right) = \sum_i [-\dot{p}_i dq_i + \dot{q}_i dp_i] - \frac{\partial L}{\partial t} dt.$$

The term on the left-hand side is just the Hamiltonian that we have defined before, so we find that

$$dH = \sum_i [-\dot{p}_i dq_i + \dot{q}_i dp_i] - \frac{\partial L}{\partial t} dt = \sum_i \left[\frac{\partial H}{\partial q_i} dq_i + \frac{\partial H}{\partial p_i} dp_i\right] + \frac{\partial H}{\partial t}$$

where the second equality holds because of the definition of the total differential of H in terms of its partial derivatives. Associating terms from both sides of the equation above yields Hamilton's equations

$$\frac{\partial H}{\partial q_j} = -\dot{p}_j, \quad \frac{\partial H}{\partial p_j} = \dot{q}_j, \quad \frac{\partial H}{\partial t} = -\frac{\partial L}{\partial t}.$$

As a Reformulation of Lagrangian Mechanics

Starting with Lagrangian mechanics, the equations of motion are based on generalised coordinates

$$\{q_j \mid j = 1, \ldots, N\}$$

and matching generalised velocities

$$\{\dot{q}_j \mid j = 1, \ldots, N\}.$$

We write the Lagrangian as

$$L(q_j, \dot{q}_j, t)$$

with the subscripted variables understood to represent all *N* variables of that type. Hamiltonian mechanics aims to replace the generalised velocity variables with generalised momentum variables, also known as *conjugate momenta*. By doing so, it is possible to handle certain systems, such as aspects of quantum mechanics, that would otherwise be even more complicated.

For each generalised velocity, there is one corresponding conjugate momentum, defined as:

$$p_j = \frac{\partial L}{\partial \dot{q}_j}.$$

In Cartesian coordinates, the generalised momenta are precisely the physical linear momenta. Incircular polar coordinates, the generalised momentum corresponding to the angular velocity is the physical angular momentum. For an arbitrary choice of generalised coordinates, it may not be possible to obtain an intuitive interpretation of the conjugate momenta.

One thing which is not too obvious in this coordinate dependent formulation is that different generalised coordinates are really nothing more than different coordinatisations of the same symplectic manifold.

The *Hamiltonian* is the Legendre transform of the Lagrangian:

$$H(q_j, p_j, t) = \sum_i \dot{q}_i p_i - L(q_j, \dot{q}_j, t).$$

If the transformation equations defining the generalised coordinates are independent of t, and the Lagrangian is a sum of products of functions (in the generalised coordinates) which are homogeneous of order 0, 1 or 2, then it can be shown that H is equal to the total energy $E = T + V$.

Each side in the definition of H produces a differential:

$$dH = \sum_i \left[\left(\frac{\partial H}{\partial q_i} \right) dq_i + \left(\frac{\partial H}{\partial p_i} \right) dp_i \right] + \left(\frac{\partial H}{\partial t} \right) dt$$

$$= \sum_i \left[\dot{q}_i \, dp_i + p_i \, d\dot{q}_i - \left(\frac{\partial L}{\partial q_i} \right) dq_i - \left(\frac{\partial L}{\partial \dot{q}_i} \right) d\dot{q}_i \right] - \left(\frac{\partial L}{\partial t} \right) dt.$$

Substituting the previous definition of the conjugate momenta into this equation and matching coefficients, we obtain the equations of motion of Hamiltonian mechanics, known as the canonical equations of Hamilton:

$$\frac{\partial H}{\partial q_j} = -\dot{p}_j, \qquad \frac{\partial H}{\partial p_j} = \dot{q}_j, \qquad \frac{\partial H}{\partial t} = -\frac{\partial L}{\partial t}.$$

Hamilton's equations are first-order differential equations, and thus easier to solve than Lagrange's equations, which are second-order. Hamilton's equations have another advantage over Lagrange's equations: if a system has a symmetry, such that a coordinate does not occur in the Hamiltonian, the corresponding momentum is conserved, and that coordinate can be ignored in the other equations of the set. Effectively, this reduces the problem from n coordinates to (n-1) coordinates. In the Lagrangian framework, of course the result that the corresponding momentum is conserved still follows immediately, but all the generalised velocities still occur in the Lagrangian - we still have to solve a system of equations in n coordinates.

The Lagrangian and Hamiltonian approaches provide the groundwork for deeper results in the theory of classical mechanics, and for formulations of quantum mechanics.

Geometry of Hamiltonian Systems

A Hamiltonian system may be understood as a fibre bundle E over time R, with the fibrse E_t, t " Rbeing the position space. The Lagrangian is thus a function on the jet bundle J over E; taking the fibre wise Legendre transform of the Lagrangian produces a function on the dual bundle over time whose fibre at t is the cotangent space T^*E_t, which comes equipped with a natural symplectic form, and this latter function is the Hamiltonian.

Generalisation to Quantum Mechanics Through Poisson Bracket

Hamilton's equations above work well for classical mechanics, but not for quantum mechanics, since the differential equations discussed assume that one can specify the exact position and momentum of the particle simultaneously at any point in time. However, the equations can be further generalised to then be extended to apply to quantum mechanics as well as to classical mechanics, through the deformation of the Poisson algebra over p and q to the algebra of Moyal brackets.

Specifically, the more general form of the Hamilton's equation reads

$$\frac{\mathrm{df}}{\mathrm{dt}} = \{\mathrm{f},\mathrm{H}\} + \frac{\partial \mathrm{f}}{\partial \mathrm{t}}$$

where f is some function of p and q, and H is the Hamiltonian. To find out the rules for evaluating a Poisson bracket without resorting to differential equations; a Poisson bracket is the name for the Lie bracket in a Poisson algebra. These Poisson brackets can then be extended to Moyal brackets comporting to an in equivalent Lie algebra, as proven by H Groenewold, and thereby describe quantum mechanical diffusion in phase space. This more algebraic approach not only permits ultimately extending probability distributions in phase space to Wigner quasi-probability distributions, but, at the mere Poisson bracket classical setting, also provides more power in helping analyse the relevant conserved quantities in a system.

Mathematical Formalism

Any smooth real-valued function *H* on a symplectic manifold can be used to define a Hamiltonian system. The function *H* is known as the Hamiltonian or the energy function. The symplectic manifold is then called the phase space. The Hamiltonian induces a special vector field on the symplectic manifold, known as the symplectic vector field.

The symplectic vector field, also called the Hamiltonian vector field, induces a Hamiltonian flow on the manifold. The integral curves of the vector field are a one-parameter family of transformations of the manifold; the parameter of the curves is commonly called the time. The time evolution is given by symplectomorphisms. By Liouville's theorem, each symplectomorphism preserves the volume form on the phase space. The collection of symplectomorphisms induced by the Hamiltonian flow is commonly called the Hamiltonian mechanics of the Hamiltonian system.

The symplectic structure induces a Poisson bracket. The Poisson bracket gives the space of functions on the manifold the structure of a Lie algebra.

Given a function *f*

$$\frac{d}{dt}f = \frac{\partial}{\partial t}f + \{f, H\}.$$

If we have a probability distribution, ρ, then (since the phase space velocity $(\dot{p}_i, \dot{q}_i)$ has zero divergence, and probability is conserved) its convective derivative can be shown to be zero and so

$$\frac{\partial}{\partial t}\rho = -\{\rho, H\}.$$

This is called Liouville's theorem. Every smooth function *G* over the symplectic manifold generates a one-parameter family of symplectomorphisms and if $\{G, H\} = 0$, then *G* is conserved and the symplectomorphisms are symmetry transformations.

A Hamiltonian may have multiple conserved quantities G_i. If the symplectic manifold has dimension $2n$ and there are n functionally independent conserved quantities G_i which are in involution (i.e., $\{G_i, G_j\} = 0$), then the Hamiltonian is Liouville integrable. The Liouville–Arnol'd theorem says that locally, any Liouville integrable Hamiltonian can be transformed via a symplectomorphism in a new Hamiltonian with the conserved quantities G_i as coordinates; the new coordinates are called *action-angle coordinates*. The transformed Hamiltonian depends only on the G_i, and hence the equations of motion have the simple form

$$\dot{G}_i = 0, \qquad \dot{\varphi}_i = F(G),$$

for some function F (Arnol'd et al., 1988). There is an entire field focusing on small deviations from integrable systems governed by the KAM theorem.

The integrability of Hamiltonian vector fields is an open question. In general, Hamiltonian systems arechaotic; concepts of measure, completeness, integrability and stability are poorly defined. At this time, the study of dynamical systems is primarily qualitative, and not a quantitative science.

Riemannian Manifolds

An important special case consists of those Hamiltonians that are quadratic forms, that is, Hamiltonians that can be written as

$$H(q,p) = \frac{1}{2}\langle p,p\rangle_q$$

where $\langle\cdot,\cdot\rangle_q$ is a smoothly varying inner product on the fibres T_q^*Q, the cotangent space to the point q in the configuration space, sometimes called a cometric. This Hamiltonian consists entirely of the kinetic term.

If one considers a Riemannian manifold or a pseudo-Riemannian manifold, the Riemannian metricinduces a linear isomorphism between the tangent and cotangent bundles. Using this isomorphism, one can define a cometric. (In coordinates, the matrix defining the cometric is the inverse of the matrix defining the metric.) The solutions to the Hamilton–Jacobi equations for this Hamiltonian are then the same as the geodesics on the manifold. In particular, the Hamiltonian flow in this case is the same thing as the geodesic flow. The existence of such solutions, and the completeness of the set of solutions, are discussed in detail in the article on geodesics.

Sub-Riemannian Manifolds

When the cometric is degenerate, then it is not invertible. In this case, one does not have a Riemannian manifold, as one does not have a metric. However, the Hamiltonian still exists. In the case where the cometric is degenerate at every point q of the configuration space manifold Q, so that therank of the cometric is less than the dimension of the manifold Q, one has a sub-Riemannian manifold.

The Hamiltonian in this case is known as a sub-Riemannian Hamiltonian. Every such Hamiltonian uniquely determines the cometric, and vice-versa. This implies that every sub-Riemannian manifold is uniquely determined by its sub-Riemannian Hamiltonian, and that the converse is true: every sub-Riemannian manifold has a unique sub-Riemannian Hamiltonian. The existence of sub-Riemannian geodesics is given by the Chow-Rashevskii theorem.

The continuous, real-valued Heisenberg group provides a simple example of a sub-Riemannian manifold. For the Heisenberg group, the Hamiltonian is given by

$$H(x,y,z,p_x,p_y,p_z)=\frac{1}{2}\left(p_x^2+p_y^2\right).$$

p_z is not involved in the Hamiltonian.

Poisson Algebras

Hamiltonian systems can be generalised in various ways. Instead of simply looking at the algebra of smooth functions over a symplectic manifold, Hamiltonian mechanics can be formulated on general commutative unital real Poisson algebras. A state is a continuous linear functional on the Poisson algebra (equipped with some suitable topology) such that for any element A of the algebra, A^2 maps to a nonnegative real number.

A further generalisation is given by Nambu dynamics.

Charged Particle in an Electromagnetic Field

A good illustration of Hamiltonian mechanics is given by the Hamiltonian of a charged particle in an electromagnetic field. In Cartesian coordinates (i.e. $q_i = x_i$), the Lagrangian of a non-relativistic classical particle in an electromagnetic field is (in SI Units):

$$L=\sum_i \tfrac{1}{2}m\dot{x}_i^2+\sum_i e\dot{x}_i A_i - e\phi,$$

where e is the electric charge of the particle (not necessarily the electron charge), ϕ is the electric scalar potential, and the A_i are the components of the magnetic vector potential (these may be modified through a gauge transformation). This is called minimal coupling.

The generalised momenta may be derived by:

$$p_j = \frac{\partial L}{\partial \dot{x}_j} = m\dot{x}_j + eA_j.$$

Rearranging, we may express the velocities in terms of the momenta, as:

$$\dot{x}_j = \frac{p_j - eA_j}{m}.$$

If we substitute the definition of the momenta, and the definitions of the velocities in terms of the momenta, into the definition of the Hamiltonian given above, and then simplify and rearrange, we get:

$$H = \sum_i \dot{x}_i p_i - L = \sum_i \frac{(p_i - eA_i)^2}{2m} + e\phi.$$

This equation is used frequently in quantum mechanics.

Relativistic Charged Particle in an Electromagnetic Field

The Lagrangian for a relativistic charged particle is given by:

$$L[t] = -mc^2\sqrt{1 - \frac{\dot{\vec{x}}[t]^2}{c^2}} - e\phi[\vec{x}[t],t] + e\dot{\vec{x}}[t]\cdot\vec{A}[\vec{x}[t],t].$$

Thus the particle's canonical (total) momentum is

$$\vec{P}[t] = \frac{\partial L[t]}{\partial \dot{\vec{x}}[t]} = \frac{m\dot{\vec{x}}[t]}{\sqrt{1 - \frac{\dot{\vec{x}}[t]^2}{c^2}}} + e\vec{A}[\vec{x}[t],t],$$

that is, the sum of the kinetic momentum and the potential momentum.

Solving for the velocity, we get

$$\dot{\vec{x}}[t] = \frac{\vec{P}[t] - e\vec{A}[\vec{x}[t],t]}{\sqrt{m^2 + \frac{1}{c^2}\left(\vec{P}[t] - e\vec{A}[\vec{x}[t],t]\right)^2}}.$$

So the Hamiltonian is

$$H[t] = \dot{\vec{x}}[t]\cdot\vec{P}[t] - L[t] = c\sqrt{m^2c^2 + \left(\vec{P}[t] - e\vec{A}[\vec{x}[t],t]\right)^2} + e\phi[\vec{x}[t]$$

From this we get the force equation (equivalent to the Euler–Lagrange equation)

$$\dot{\vec{P}} = -\frac{\partial H}{\partial \vec{x}} = e(\vec{\nabla}\vec{A}) \cdot \dot{\vec{x}} - e\vec{\nabla}\phi$$

from which one can derive

$$\frac{d}{dt}\left(\frac{m\dot{\vec{x}}}{\sqrt{1-\frac{\dot{\vec{x}}^2}{c^2}}}\right) = e\vec{E} + e\dot{\vec{x}} \times \vec{B}.$$

An equivalent expression for the Hamiltonian as function of the relativistic (kinetic) momentum, $\vec{p} = \gamma m\dot{\vec{x}}[t]$, is

$$H[t] = \dot{\vec{x}}[t] \cdot \vec{p}[t] + \frac{mc^2}{\gamma} + e\phi[\vec{x}[t],t] = \gamma mc^2 + e\phi[\vec{x}[t],t] = E + V.$$

This has the advantage that $\vec{p}$ can be measured experimentally whereas $\vec{p}$ cannot. Notice that the Hamiltonian (total energy) can be viewed as the sum of the relativistic energy (kinetic+rest), $E = \gamma mc^2$, plus the potential energy, $V = e\phi$.

Chapter 6

Fundamental Concepts of Mechanics

Space

Space is the boundless, three-dimensional extent in which objects and events occur and have relative position and direction. Physical space is often conceived in three linear dimensions, although modern physicists usually consider it, with time, to be part of a boundless four-dimensional continuum known as spacetime. In mathematics one examines "spaces" with different numbers of dimensions and with different underlying structures. The concept of space is considered to be of fundamental importance to an understanding of the physical universe although disagreement continues between philosophers over whether it is itself an entity, a relationship between entities, or part of a conceptual framework.

Debates concerning the nature, essence and the mode of existence of space date back to antiquity; namely, to treatises like the *Timaeus* of Plato, or Socrates in his reflections on what the Greeks called khora (i.e. "space"), or in the *Physics* of Aristotle (Book IV, Delta) in the definition of *topos* (i.e. place), or even in the later "geometrical conception of place" as "space *qua* extension" in the *Discourse on Place* (*Qawl fi al-Makan*) of the 11th century Arab polymath Alhazen. Many of these classical philosophical questions were discussed in the Renaissance and then reformulated in the 17th century, particularly during the early development of classical mechanics. In Isaac Newton's view, space was absolute - in the sense that it existed permanently and independently of whether there were any matter in the space. Other natural philosophers, notably Gottfried Leibniz, thought instead that space was a collection of relations between objects, given by their distance and direction from one another. In the 18th century, the philosopher and theologian George Berkeley attempted to refute the "visibility of spatial depth" in his *Essay Towards a New Theory of Vision*. Later, the metaphysician Immanuel

Kant said neither space nor time can be empirically perceived, they are elements of a systematic framework that humans use to structure all experiences. Kant referred to "space" in his *Critique of Pure Reason* as being: a subjective "pure *a priori* form of intuition", hence it is an unavoidable contribution of our human faculties.

In the 19th and 20th centuries mathematicians began to examine non-Euclidean geometries, in which space can be said to be *curved*, rather than *flat*. According to Albert Einstein's theory of general relativity, space around gravitational fields deviates from Euclidean space. Experimental tests of general relativity have confirmed that non-Euclidean space provides a better model for the shape of space.

Time

Figure 1: The flow of sand in an hourglass can be used to keep track of elapsed time. It also concretely represents the present as being between the past and the future.

Time is a part of the measuring system used to sequence events, to compare the durations of events and the intervals between them, and to quantify rates of change such as the motions of objects. The temporal position of events with respect to the transitory present is continually changing; events happen, then are located further and further in the past. Time has been a major subject of religion, philosophy, and science, but defining it in a non-controversial manner applicable to all fields of study has consistently eluded the greatest scholars. A simple definition states that "time is what clocks measure".

Time is one of the seven fundamental physical quantities in the International System of Units. Time is used to define other quantities — such as velocity — so defining time in terms of such quantities would result in circularity of definition. An operational definition of time, wherein one says that observing a certain number of repetitions of one or another standard cyclical event (such as the passage of a free-swinging pendulum) constitutes one standard unit such as the second, is highly useful in the conduct of both advanced experiments and everyday affairs of life. The operational definition leaves aside the question whether there is something called time, apart from the counting activity just mentioned, that flows and that can be measured. Investigations of a single continuum called spacetime bring questions about space into questions about time, questions that have their roots in the works of early students of natural philosophy.

Two contrasting viewpoints on time divide many prominent philosophers. One view is that time is part of the fundamental structure of the universe, a dimension in which events occur in sequence. Sir Isaac Newton subscribed to this realist view, and hence it is sometimes referred to as Newtonian time. Time travel, in this view, becomes a possibility as other "times" persist like frames of a film strip, spread out across the time line. The opposing view is that *time* does not refer to any kind of "container" that events and objects "move through", nor to any entity that "flows", but that it is instead part of a fundamental intellectual structure (together with space and number) within which humans sequence and compare events. This second view, in the tradition of Gottfried Leibniz and Immanuel Kant, holds that *time* is neither an event nor a thing, and thus is not itself measurable nor can it be travelled.

Temporal measurement has occupied scientists and technologists, and was a prime motivation in navigation and astronomy. Periodic events and periodic motion have long served as standards for units of time. Examples include the apparent motion of the sun across the sky, the phases of the moon, the swing of a pendulum, and the beat of a heart. Currently, the international unit of time, the second, is defined in terms of radiation emitted by caesium atoms. Time is also of significant social importance, having economic value ("time is money") as well as personal value, due to an awareness of the limited time in each day and in human life spans.

Ray Cummings, an early writer of science fiction, wrote in 1922, "Time... is what keeps everything from happening at once", a sentence repeated by scientists such as C. J. Overbeck, and John Archibald Wheeler.

Velocity

In physics, velocity is speed in a given direction. Speed describes only how fast an object is moving, whereas velocity gives both the speed and direction of the object's motion. To have a constant velocity, an object must have a constant speed and motion in a constant direction. Constant direction, typically constrains the object to motion in a straight path. A car moving at a constant 20 kilometres per hour in a circular path does not have a constant velocity. The rate of change in velocity is acceleration. Velocity is a vector physical quantity; both magnitude and direction are required to define it. Thescalar absolute value (magnitude) of velocity is speed, a quantity that is measured in metres per second (m/s or ms^{-1}) when using the SI (metric) system.

For example, -5 metres per second" is a scalar and not a vector, whereas -5 metres per second east" is a vector. The *average* velocity v of an object moving through a displacement (Δx) during a time interval (Δt) is described by the formula:

$$\bar{v} = \frac{\Delta x}{\Delta t}.$$

The rate of change of velocity is acceleration – how an object's speed or direction of travel changes over time, and how it is changing at a particular point in time.

Equation of Motion

The velocity vector v of an object that has positions x(t) at time t and x($t + \Delta t$) at time $t + \Delta t$, can be computed as the derivative of position:

$$v = \lim_{\Delta t \to 0} \frac{x(t + \Delta t) - x(t)}{\Delta t} = \frac{dx}{dt}.$$

Average velocity magnitudes always smaller than or equal to average speed of a given particle. Instantaneous velocity is always tangential to trajectory. Slope of tangent of position or displacement time graph is instantaneous velocity and its slope of chord is average velocity.

The equation for an object's velocity can be obtained mathematically by evaluating the integral of the equation for its acceleration beginning from some initial period time t_0 to some point in time later t_n.

The final velocity v of an object which starts with velocity u and then accelerates at constant acceleration a for a period of time Δt is:

$$v = u + a\Delta t.$$

The average velocity of an object undergoing constant acceleration is $\frac{(u+v)}{2}$, where u is the initial velocity and v is the final velocity. To find the position, x, of such an accelerating object during a time interval, Δt, then:

$$\Delta x = \frac{(u+v)}{2}\Delta t.$$

When only the object's initial velocity is known, the expression,

$$\Delta x = u\Delta t + \frac{1}{2}a\Delta t^2,$$

can be used.

This can be expanded to give the position at any time t in the following way:

$$x(t) = x(0) + \Delta x = x(0) + u\Delta t + \frac{1}{2}a\Delta t^2,$$

These basic equations for final velocity and position can be combined to form an equation that is independent of time, also known as Torricelli's equation:

$$v^2 = u^2 + 2a\Delta x.$$

The above equations are valid for both Newtonian mechanics and special relativity. Where Newtonian mechanics and special relativity differ is in how different observers would describe the same situation. In particular, in Newtonian mechanics, all observers agree on the value of t and the transformation rules for position create a situation in which all non-accelerating observers would describe the acceleration of an object with the same values. Neither is true for special relativity. In other words only relative velocity can be calculated.

In Newtonian mechanics, the kinetic energy (energy of motion), E_K, of a moving object is linear with both its mass and the square of its velocity:

$$E_K = \frac{1}{2}mv^2.$$

The kinetic energy is a scalar quantity.

Escape velocity is the minimum velocity a body must have in order to escape from the gravitational field of the earth. To escape from the Earth's gravitational field an object must have greater kinetic energy than its gravitational potential energy. The value of the escape velocity from the Earth's surface is approximately 11100 m/s.

Speed

In kinematics, the speed of an object is the magnitude of its velocity (the rate of change of its position); it is thus a scalar quantity. The average speed of an object in an interval of time is the distance travelled by the object divided by the duration of the interval; the instantaneous speed is the limit of the average speed as the duration of the time interval approaches zero .

Like velocity, speed has the dimensions of a length divided by a time; the SI unit of speed is the metre per second, but the most usual unit of speed in everyday usage is the kilometre per hour or, in the USA and the UK, miles per hour. For air and marine travel the knot is commonly used.

The fastest possible speed at which energy or information can travel, according to special relativity, is the speed of light in vacuum c = 299,792,458 metres per second, approximately 1079 million kilometres per hour (671,000,000 mph). Matter cannot quite reach the speed of light, as this would require an infinite amount of energy.

Definition

The speed v is defined as the magnitude of the velocity v, that is the derivative of the position r with respect to time:

$$\mathrm{v} = |\mathrm{v}| = |\dot{\mathrm{r}}| = \left|\frac{\mathrm{dr}}{\mathrm{dt}}\right|.$$

If s is the length of the path travelled until time t, the speed equals the time derivative of s:

$$\mathrm{v} = \frac{\mathrm{ds}}{\mathrm{dt}}.$$

In the special case where the velocity is constant (that is, constant speed in a straight line) this can be simplified to $v=s/t$. The average speed over a finite time interval is the total distance travelled divided by the time duration.

Expressed in graphical language, the slope of a tangent line of a distance-time graph is the instantaneous speed, and the slope of a chord line of distance-time graph is the average speed over the time interval between the ends of the chord.

Units

Units of speed include:

- metres per second (symbol m s^{-1} or m/s), the SI derived unit;
- kilometres per hour (symbol km/h);
- miles per hour (symbol mph);
- knots (nautical miles per hour, symbol kn or kt);
- feet per second (symbol fps or ft/s);
- Mach number (dimensionless), speed divided by the speed of sound;
- in natural units (dimensionless), speed divided by the speed of light in vacuum (symbol c = 299,792,458 m/s).

Conversions between common units of speed

	m/s	*Km/h*	*mph*	*knot*	*ft/s*
1 m/s =	1	3.6	2.236936	1.943844	3.280840
1 km/h =	0.277778	1	0.621371	0.539957	0.911344
1 mph =	0.44704	1.609344	1	0.868976	1.466667
1 knot =	0.514444	1.852	1.150779	1	1.687810
1 ft/s =	0.3048	1.09728	0.681818	0.592484	1

(Values in bold face are exact.)

Mass

Mass can be defined as a quantitive measure of the resistance an object has to change in its velocity. Inphysics, mass commonly refers to any of the following three properties of matter, which have been shown experimentally to be equivalent:

- inertial mass
- active gravitational mass
- passive gravitational mass.

Mass must be distinguished from matter in physics, because matter is a poorly-defined concept, and although all types of agreed-upon matter exhibit mass, it is also the case that many types of energy which are not matter—such as potential energy, kinetic energy, and trapped electromagnetic radiation (photons)—also exhibit mass. Thus, all matter has the property of mass, but not all mass is associated with identifiable matter.

In everyday usage, "mass" is often used interchangeably with *weight*, and the units of weight are often taken to be kilograms (for instance, a person may state that their weight is 75kg). In scientific use, however, the two terms refer to different, yet related, properties of matter. Weight can be zero if no gravitational force is acting but mass can never be zero. The inertial mass of an object determines its acceleration in the presence of an applied force. According to Newton's second law of motion, if a body of fixed mass M is subjected to a force F, its acceleration α is given by F/M.

A body's mass also determines the degree to which it generates or is affected by a gravitational field. If a first body of mass M_A is placed at a distance r from a second body of mass M_B, each body experiences an attractive force F whose magnitude is

$$F = \frac{GM_A M_B}{r^2}$$

where G is the universal constant of gravitation, equal to 6.67×10^{-11} N m^2kg^{-2}. This is sometimes referred to as gravitational mass (when a distinction is necessary, M is used to denote the active gravitational mass and m the passive gravitational mass). Repeated experiments since the 17th century have demonstrated that inertial and gravitational mass are equivalent; this is entailed in the equivalence principle of general relativity.

Special relativity shows that rest mass (or invariant mass) and rest energy are essentially equivalent, via the well-known relationship ($E=mc^2$). This same equation also connects relativistic mass and "relativistic energy" (total system energy). These are concepts that are related to their "rest" counterparts, but they do not have the same value, in systems where there is a net momentum. In order to deduce any of these four quantities from any of the others, in any system which has a net momentum, an equation that takes momentum into account is needed.

Mass (so long as the type and definition of mass is agreed upon) is a conserved quantity over time. From the viewpoint of any single unaccelerated observer, mass can neither be created or destroyed, and special relativity does not change this understanding (though different observers may not agree on how much mass is present, all agree that the amount does not change over time). However, relativity adds the fact that all types of energy have an associated mass, and this mass is added to systems when energy is added, and the associated mass is subtracted from systems when the energy leaves. In such cases, the energy leaving or entering the system carries the added or missing mass with it, since this energy itself has mass. Thus, mass remains conserved when the location of all mass is taken into account.

On the surface of the Earth, the weight *W* of an object is related to its mass *m* by

$$W = mg,$$

where *g* is the Earth's gravitational field strength, equal to about 9.81 m s^{-2}. An object's weight depends on its environment, while its mass does not: an object with a mass of 50 kilograms weighs 491 newtons on the surface of the Earth; on the surface of the Moon, the same object still has a mass of 50 kilograms but weighs only 81.5 newtons.

Units of Mass

In the International System of Units (SI), mass is measured in kilograms (kg). The gram (g) is $^{1}D_{1000}$ of a kilogram.

Other units are accepted for use in SI:

- The tonne (t) is equal to 1000 kg.
- The electronvolt (eV) is primarily a unit of energy, but because of the mass-energy equivalence it can also function as a unit of mass. In this context it is denoted eV/c^2, or simply as eV. The electronvolt is common in particle physics.
- The atomic mass unit (u) is defined so that a single carbon-12 atom has a mass of 12 u; 1 u is approximately 1.66×10^{-27} kg. The atomic mass unit is convenient for expressing the masses of atoms and molecules.

Outside the SI system, a variety of different mass units are used, depending on context, such as the slug (sl), the pound (lb), the Planck mass (m_P), and the solar mass

In normal situations, the weight of an object is proportional to its mass, which usually makes it unproblematic to use the same unit for both concepts. However, the distinction between mass and weight becomes important for measurements with a precision better than a few percent (because of slight differences in the strength of the Earth's gravitational field at different places), and for places far from the surface of the Earth, such as in space or on other planets. A mass can sometimes be expressed in terms of length. The mass of a very small particle may be identified with its inverse Compton wavelength (1 $cm^{-1} \approx 3.52\times10^{-41}$ kg). The mass of a very large star or black hole may be identified with its Schwarzschild radius (1 cm $\approx 6.73\times10^{24}$ kg).

Acceleration

In physics, acceleration is the rate of change of velocity with time. In one dimension, acceleration is the rate at which something speeds up or slows down. However, since velocity is a vector, acceleration describes the rate of change of both the magnitude and the direction of velocity. Acceleration has the dimensions $L\,T^{-2}$. InSI units, acceleration is measured in metres per second squared (m/s^2). (Negative acceleration i.e. retardation, also has the same dimensions/units.) Proper acceleration, the acceleration of a body relative to a free-fall condition, is measured by an instrument called an accelerometre. In common speech, the term *acceleration* is used for an increase in speed (the magnitude of velocity); a decrease in speed is called *deceleration*. In physics, a change in the direction of velocity also is an acceleration: for rotary motion, the change in direction of velocity results in *centripetal (toward the centre) acceleration*; whereas the rate of change of speed is a *tangential acceleration*.

In classical mechanics, for a body with constant mass, the acceleration of the body is proportional to the net force acting on it (Newton's second law):

$$F = ma \rightarrow a = F/m$$

where F is the resultant force acting on the body, m is the mass of the body, and a is its acceleration.

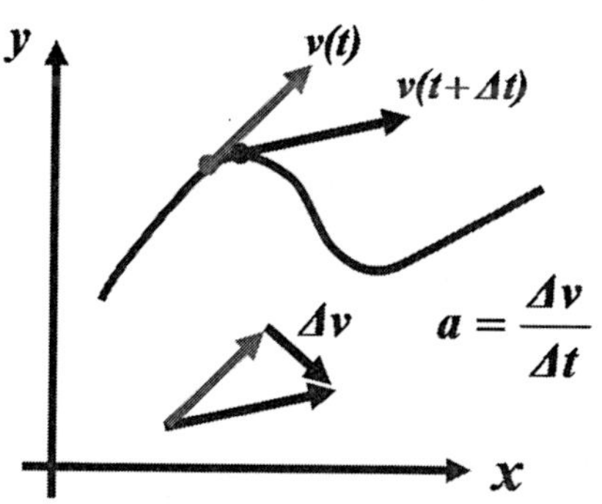

Figure 2: Acceleration is the rate of change of velocity. At any point on a trajectory, the magnitude of the acceleration is given by the rate of change of velocity in both magnitude and direction at that point. The true acceleration at time t is found in the limit as time interval $\Delta t \rightarrow 0$.

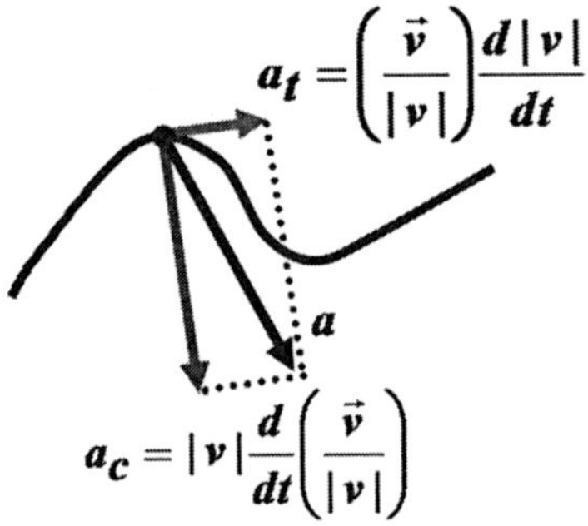

Figure 3: Components of acceleration for a planar curved motion. The tangential component a_t is due to the change in speed of traversal, and points along the curve in the direction of the velocity vector. The centripetal component a_c is due to the change in direction of the velocity vector and is normal to the trajectory, pointing toward the centre of curvature of the path.

Average acceleration is the change in velocity (Δv) divided by the change in time (Δt). Instantaneous acceleration is the acceleration at a specific point in time which is for a very short interval of time as Δt approaches zero.

Tangential and Centripetal Acceleration

The velocity of a particle moving on a curved path as a function of time can be written as:

$$v(t) = v(t)\frac{v(t)}{v(t)} = v(t)u_t(t),$$

with $v(t)$ equal to the speed of travel along the path, and

$$u_t = \frac{v(t)}{v(t)},$$

a unit vector tangent to the path pointing in the direction of motion at the chosen moment in time. Taking into account both the changing speed $v(t)$ and the changing direction of u_t, the acceleration of a particle moving on a curved path on a planar surface can be written using the chain rule of differentiation and the derivative of the product of two functions of time as:

$$\begin{aligned} a &= \frac{dv}{dt} \\ &= \frac{dv}{dt}u_t + v(t)\frac{du_t}{dt} \\ &= \frac{dv}{dt}u_t + \frac{v^2}{R}u_n, \end{aligned}$$

where u_n is the unit (inward) normal vector to the particle's trajectory, and R is its instantaneous radius of curvature based upon the osculating circle at time t. These components are called the tangential acceleration and the radial acceleration or centripetal acceleration.

Extension of this approach to three-dimensional space curves that cannot be contained on a planar surface leads to the Frenet–Serret formulas.

Gravitation

Gravitation, or gravity, is a natural phenomenon by which physical bodies attract with a force proportional to their mass. Gravitation is most familiar as the agent that gives weight to objects with mass and causes them to fall to the ground when dropped. Gravitation causes dispersed matter to coalesce, and coalesced matter to remain intact, thus accounting for the existence of the Earth, the Sun, and most of the macroscopic objects in the universe.

Gravitation is responsible for keeping the Earth and the other planets in their orbits around the Sun; for keeping the Moon in its orbit around the Earth; for the formation of tides; for natural convection, by which fluid flow occurs under the influence of a density gradient and gravity; for heating the interiors of forming stars and planets to very high temperatures; and for various other phenomena observed on Earth.

Gravitation is one of the four fundamental interactions of nature, along with electromagnetism, and the nuclear strong force and weak force. Modern physics describes gravitation using the general theory of relativity by Einstein, in which it is a consequence of the curvature of spacetime governing the motion of inertial objects. The simpler Newton's law of universal gravitation provides an accurate approximation for most physical situations.

History of Gravitational Theory

Scientific Revolution: Modern work on gravitational theory began with the work of Galileo Galilei in the late 16th and early 17th centuries. In his famous (though possibly apocryphal) experiment dropping balls from the Tower of Pisa, and later with careful measurements of balls rolling down inclines, Galileo showed that gravitation accelerates all objects at the same rate. This was a major departure from Aristotle's belief that heavier objects accelerate faster. Galileo correctly postulated air resistance as the reason that lighter objects may fall more slowly in an atmosphere. Galileo's work set the stage for the formulation of Newton's theory of gravity.

Newton's Theory of Gravitation

In 1687, English mathematician Sir Isaac Newton published *Principia*, which hypothesizes the inverse-square law of universal gravitation. In his own words, "I deduced that the forces which keep the planets in their orbs must [be] reciprocally as the squares of their distances from the centres about which they revolve: and thereby compared the force requisite to keep the Moon in her Orb with the force of gravity at the surface of the Earth; and found them answer pretty nearly."

Newton's theory enjoyed its greatest success when it was used to predict the existence of Neptune based on motions of Uranus that could not be accounted for by the actions of the other planets. Calculations by both John Couch Adams and Urbain Le Verrier predicted the general position of the planet, and Le Verrier's calculations are what led Johann Gottfried Galle to the discovery of Neptune. A discrepancy in Mercury's orbit pointed out flaws in Newton's theory. By the end of the 19th century, it was known that its orbit showed slight perturbations that could not be accounted for entirely under Newton's theory, but all searches for another perturbing body (such as a planet orbiting the Sun even closer than Mercury) had been fruitless. The issue was resolved in 1915 by Albert Einstein's new theory of general relativity, which accounted for the small discrepancy in Mercury's orbit.

Although Newton's theory has been superseded, most modern non-relativistic gravitational calculations are still made using Newton's theory because it is a much simpler theory to work with than general relativity, and gives sufficiently accurate results for most applications involving sufficiently small masses, speeds and energies.

Equivalence Principle

The equivalence principle, explored by a succession of researchers including Galileo, Loránd Eötvös, and Einstein, expresses the idea that all objects fall in the same way. The simplest way to test the weak equivalence principle is to drop two objects of different masses or compositions in a vacuum, and see if they hit the ground at the same time. These experiments demonstrate that all objects fall at the same rate when friction (including air resistance) is negligible. More sophisticated tests use a torsion balance of a type invented by Eötvös.

Satellite experiments, for example STEP, are planned for more accurate experiments in space.

Formulations of the equivalence principle include:

- The weak equivalence principle: *The trajectory of a point mass in a gravitational field depends only on its initial position and velocity, and is independent of its composition.*
- The Einsteinian equivalence principle: *The outcome of any local non-gravitational experiment in a freely falling laboratory is independent of the velocity of the laboratory and its location in spacetime.*
- The strong equivalence principle requiring both of the above.

The equivalence principle can be used to make physical deductions about the gravitational constant, the geometrical nature of gravity, the possibility of a fifth force, and the validity of concepts such as general relativity and Brans-Dicke theory.

General Relativity

In general relativity, the effects of gravitation are ascribed to spacetime curvature instead of a force. The starting point for general relativity is the equivalence principle, which equates free fall with inertial motion, and describes free-falling inertial objects as being accelerated relative to non-inertial observers on the ground. In Newtonian physics, however, no such acceleration can occur unless at least one of the objects is being operated on by a force.

Einstein proposed that spacetime is curved by matter, and that free-falling objects are moving along locally straight paths in curved spacetime. These straight paths are called geodesics. Like Newton's first law of motion, Einstein's theory states that if a force is applied on an object, it would deviate from a geodesic. For instance, we are no longer following geodesics while standing because the mechanical resistance of the Earth exerts an upward force on us, and we are non-inertial on the ground as a result. This explains why moving along the geodesics in spacetime is considered inertial.

Einstein discovered the field equations of general relativity, which relate the presence of matter and the curvature of spacetime and are named after him. The Einstein field equations are a set of 10 simultaneous, non-linear, differential equations. The solutions of the field equations are the components of the metric tensor of spacetime. A metric tensor describes a geometry of spacetime. The geodesic paths for a spacetime are calculated from the metric tensor.

Notable solutions of the Einstein field equations include:

- The Schwarzschild solution, which describes spacetime surrounding a spherically symmetricnon-rotating uncharged massive object. For compact enough objects, this solution generated a black hole with a central singularity. For radial distances from the centre which are much greater than the Schwarzschild radius, the accelerations predicted by the Schwarzschild solution are practically identical to those predicted by Newton's theory of gravity.
- The Reissner-Nordström solution, in which the central object has an electrical charge. For charges with a geometrised length which are less than the geometrised length of the mass of the object, this solution produces black holes with two event horizons.
- The Kerr solution for rotating massive objects. This solution also produces black holes with multiple event horizons.
- The Kerr-Newman solution for charged, rotating massive objects. This solution also produces black holes with multiple event horizons.
- The cosmological Friedmann-Lemaitre-Robertson-Walker solution, which predicts the expansion of the universe.

The tests of general relativity included the following:

- General relativity accounts for the anomalous perihelion precession of Mercury.
- The prediction that time runs slower at lower potentials has been confirmed by the Pound–Rebka experiment, the Hafele–Keating experiment, and the GPS.
- The prediction of the deflection of light was first confirmed by Arthur Stanley Eddington from his observations during the Solar eclipse of May 29, 1919. Eddington measured starlight deflections twice those predicted by Newtonian corpuscular theory, in accordance with the predictions of general relativity. However his interpretation of the results was later disputed. More recent tests using radio interferometric measurements of quasars passing behind the Sun have more accurately and consistently confirmed the deflection of light to the degree predicted by general relativity.

- The time delay of light passing close to a massive object was first identified by Irwin I. Shapiro in 1964 in interplanetary spacecraft signals.
- Gravitational radiation has been indirectly confirmed through studies of binary pulsars.
- Alexander Friedmann in 1922 found that Einstein equations have non-stationary solutions (even in the presence of the cosmological constant). In 1927 Georges Lemaître showed that static solutions of the Einstein equations, which are possible in the presence of the cosmological constant, are unstable, and therefore the static universe envisioned by Einstein could not exist. Later, in 1931, Einstein himself agreed with the results of Friedmann and Lemaître. Thus general relativity predicted that the Universe had to be non-static—it had to either expand or contract. The expansion of the universe discovered by Edwin Hubble in 1929 confirmed this prediction.
- The theory's prediction of frame dragging was consistent with the recent Gravity Probe B results.
- General relativity predicts that the light should lose its energy when travelling away from the massive bodies. The group of Radek Wojtak of the Niels Bohr Institute at the University of Copenhagen collected the data from 8000 galaxy clusters and found that the light coming from the cluster centres tended to be redshifted compared to the cluster egdes, confirming the energy loss due to the gravity.

Force

In physics, a force is any influence that causes an object to undergo a change in speed, a change in direction, or a change in shape. In other words, a force is that which can cause an object with mass to change its velocity (which includes to begin moving from a state of rest), i.e., to accelerate, or which can cause a flexible object to deform. Force can also be described by intuitive concepts such as a push or pull. A force has both magnitude and direction, making it a vector quantity. Newton's second law, F=ma, was originally formulated in slightly different, but equivalent terms: the original version states that the net force acting upon an object is equal to the rate at which its momentum changes.

Related concepts to force include: thrust, which increases the velocity of an object; drag, which decreases the velocity of an object;

and torque which produces changes in rotational speed of an object. Forces which do not act uniformly on all parts of a body will also cause mechanical stresses, a technical term for influences which cause deformation of matter. While mechanical stress can remain embedded in a solid object, gradually deforming it, mechanical stress in a fluid determines changes in its pressure and volume.

Impulse

In classical mechanics, an impulse (abbreviated I or J) is defined as the integral of a force with respect to time. When a force is applied to a rigid body it changes the momentum of that body. A small force applied for a long time can produce the same momentum change as a large force applied briefly, because it is the product of the force and the time for which it is applied that is important. The impulse is equal to the change of momentum.

Mathematical Derivation in the Case of an Object of Constant Mass

Impulse I produced from time t_1 to t_2 is defined to be

$$I = \int_{t_1}^{t_2} F dt$$

where F is the force applied from t_1 to t_2.

From Newton's second law, force is related to momentum p by

$$F = \frac{dp}{dt}.$$

Therefore

$$\begin{aligned} I &= \int_{t_1}^{t_2} \frac{dp}{dt} dt \\ &= \int_{p_1}^{p_2} dp \\ &= \Delta p, \end{aligned}$$

where Δp is the change in momentum from time t_1 to t_2. This is often called the impulse-momentum theorem.

As a result, an impulse may also be regarded as the change in momentum of an object to which a force is applied. The impulse may be expressed in a simpler form when both the force and the mass are constant:

$$I = F\Delta t = m\Delta v = \Delta p$$

where

F is the *constant* total net force applied,

Δt is the time interval over which the force is applied,

m is the *constant* mass of the object,

Δv is the change in velocity produced by the force in the considered time interval, and

Δp is the change in linear momentum.

It is often the case that not just one but both of these two quantities vary.

In the technical sense, impulse is a physical quantity, not an event or force. The term "impulse" is also used to refer to a fast-acting force. This type of impulse is often *idealised* so that the change in momentum produced by the force happens with no change in time. This sort of change is a step change, and is not physically possible. This is a useful model for computing the effects of ideal collisions (such as in game physics engines).

Impulse has the same units (in the International System of Units, kg·m/s = N·s) and dimensions (MLT^{-1}) as momentum.

Impulse can be calculated using the equation

$$F\Delta t = \Delta p = mv_1 - mv_0$$

where

F is the *constant* total net force applied,

t is the time interval over which the force is applied,

m is the *constant* mass of the object,

v_1 is the final velocity of the object at the end of the time interval, and

v_0 is the initial velocity of the object when the time interval begins.

Torque

Torque, moment or moment of force, is the tendency of a force to rotate an object about an axis, fulcrum, or pivot. Just as a force is a push or a pull, a torque can be thought of as a twist.

Loosely speaking, torque is a measure of the turning force on an object such as a bolt or a flywheel. For example, pushing or pulling the handle of a wrench connected to a nut or bolt produces a torque (turning force) that loosens or tightens the nut or bolt.

The symbol for torque is typically τ, the Greek letter *tau*. When it is called moment, it is commonly denoted *M*.

The magnitude of torque depends on three quantities: the force applied, the length of the *lever arm* connecting the axis to the point of force application, and the angle between the force vector and the lever arm. In symbols:

$$\tau = r \times F$$

$$\tau = rF \sin \theta$$

where

τ is the torque vector and τ is the magnitude of the torque,

r is the displacement vector (a vector from the point from which torque is measured to the point where force is applied), and *r* is the length (or magnitude) of the lever arm vector,

F is the force vector, and *F* is the magnitude of the force,

× denotes the cross product,

θ is the angle between the force vector and the lever arm vector.

The length of the lever arm is particularly important; choosing this length appropriately lies behind the operation of levers, pulleys, gears, and most other simple machines involving a mechanical advantage.

The SI unit for torque is the newton metre (N·m).

Terminology

This article follows US physics terminology by using the word *torque*. In the UK and in US mechanical engineering, this is called *moment*.

In US Mechanical Engineering, the term *torque* means 'the resultant moment of a Couple', and (unlike in US physics), the terms *torque* and *moment* are not interchangeable. *Torque* is defined mathematically as the rate of change of angular momentum of an object. The definition of torque states that one or both of the angular velocity or the moment of inertia of an object are changing. And *moment* is the general term used for the tendency of one or more applied forces to rotate an object about an axis, but not necessarily to change the angular momentum of the object (the concept which in physics is called torque).

For example, a rotational force applied to a shaft causing acceleration, such as a drill bit accelerating from rest, the resulting moment is called a *torque*. By contrast, a lateral force on a beam produces a moment (called a bending moment), but since the angular momentum of the beam is not changing, this bending moment is not called a *torque*. Similarly with any force couple on an object that has no change to its angular momentum, such moment is also not called a *torque*.

This article follows the US physics terminology by calling all moments by the term *torque*, whether or not they cause the angular momentum of an object to change.

History

The concept of torque, also called moment or couple, originated with the studies of Archimedes on levers. The rotational analogues of force, mass, and acceleration are torque, moment of inertia and angular acceleration, respectively.

Definition and Relation to Angular Momentum

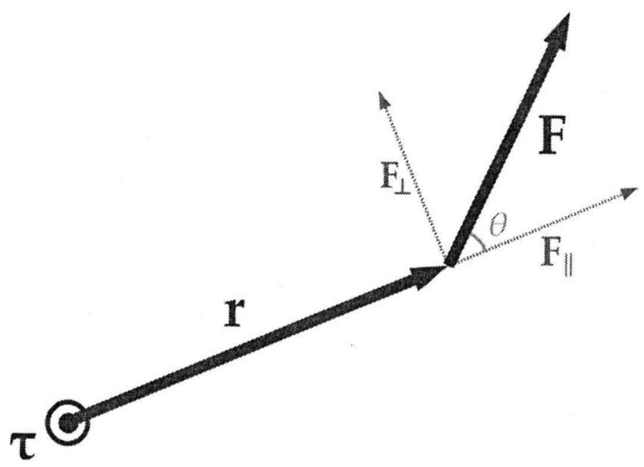

Figure 4: A particle is located at position r relative to its axis of rotation. When a force F is applied to the particle, only the perpendicular component $F_{\perp}$ produces a torque. This torque $\tau = r \times F$ has magnitude $\tau = |r|$ |F¥"| = |r| |F| $\sin\theta$ and is directed outward from the page.

Torque is defined about a point not specifically about axis as mentioned in several books.

A force applied at a right angle to a lever multiplied by its distance from the lever's fulcrum (the length of the lever arm) is its torque. A force of three newtons applied two metres from the fulcrum, for example, exerts the same torque as a force of one newton applied six metres from the fulcrum. The direction of the torque can be determined by using the right hand grip rule: if the fingers of the right hand curl in the direction of rotation and the thumb points along the axis of rotation, then the thumb also points in the direction of the torque.

More generally, the torque on a particle (which has the position r in some reference frame) can be defined as the cross product:

$$\tau = r \times F,$$

where r is the particle's position vector relative to the fulcrum, and F is the force acting on the particle. The magnitude τ of the torque is given by

$$\tau = rF\sin\theta,$$

where r is the distance from the axis of rotation to the particle, F is the magnitude of the force applied, and θ is the angle between the position and force vectors. Alternatively,

$$\tau = rF_{\perp},$$

where $F_{\perp}$ is the amount of force directed perpendicularly to the position of the particle. Any force directed parallel to the particle's position vector does not produce a torque.

It follows from the properties of the cross product that the torque vector is perpendicular to both the position and force vectors. It points along the axis of rotation, and its direction is determined by the right-hand rule.

The unbalanced torque on a body along axis of rotation determines the rate of change of the body's angular momentum,

$$\tau = \frac{dL}{dt}$$

where L is the angular momentum vector and t is time. If multiple torques are acting on the body, it is instead the net torque which determines the rate of change of the angular momentum:

$$\tau_1 + \cdots + \tau_n = \tau_{net} = \frac{dL}{dt}.$$

For rotation about a fixed axis,

$$L = I\omega,$$

where I is the moment of inertia and ω is the angular velocity. It follows that where α is the angular acceleration of the body, measured in rad·s^{-2}. This equation has the limitation that the torque equation is to be only written about instantaneous axis of rotation or centre of mass for any type of motion - either motion is pure translation, pure rotation or mixed motion. I = Moment of inertia about point about which torque is written (either about instantaneous axis of rotation or centre of mass only). If body is in translatory equilibrium then the torque equation is same about all points in the plane of motion.

Proof of the Equivalence of Definitions

The definition of angular momentum for a single particle is:

$$L = r \times p$$

where "×" indicates the vector cross product, p is the particle's linear momentum, and r is the displacement vector from the origin (the origin is assumed to be a fixed location anywhere in space). The time-derivative of this is:

$$\frac{dL}{dt} = r \times \frac{dp}{dt} + \frac{dr}{dt} \times p.$$

This result can easily be proven by splitting the vectors into components and applying the product rule. Now using the definition of linear momentum p = mv (if mass is constant) and the definition of velocity $\frac{dr}{dt} = v$

$$\frac{dL}{dt} = r \times m\frac{dv}{dt} + v \times mv.$$

The cross product of any vector with itself is zero, so the second term vanishes. Hence with the definition of force F = ma (Newton's 2nd law),

$$\frac{dL}{dt} = r \times F.$$

Then by definition, torque $\tau = r \times F$.

If multiple forces are applied, Newton's second law instead reads $F_{net} = ma$, and it follows that

$$\frac{dL}{dt} = r \times F_{net} = \tau_{net}.$$

The proof relies on the assumption that mass is constant; this is valid only in non-relativistic systems in which no mass is being ejected.

Units

Torque has dimensions of force times distance. Official SI literature suggests using the unit *newton metre* (N·m) or the unit *joule per radian*. The unit *newton metre* is properly denoted *N·m* or *N m*. This avoids ambiguity with *mN*, millinewtons.

The joule, which is the SI unit for energy or work, is dimensionally equivalent to a newton metre, but it is not used for torque. Energy and torque are entirely different concepts, so the practice of using different unit names for them helps avoid mistakes and misunderstandings. The dimensional equivalence of these units, of course, is not simply a coincidence: A torque of 1 N·m applied through a full revolution will require an energy of exactly 2π joules. Mathematically,

$$E = \tau\theta$$

where E is the energy, τ is magnitude of the torque, and θ is the angle moved (in radians). This equation motivates the alternate unit name *joules per radian*.

In British unit, "pound-force-feet" (lbf x ft), "foot-pounds-force", "inch-pounds-force", "ounce-force-inches" (oz x in) are used, and other non-SI units of torque includes "metre-kilograms-force". For all these units, the word "force" is often left out, for example abbreviating "pound-force-foot" to simply "pound-foot" (in this case, it would be implicit that the "pound" is pound-force and not pound-mass).

Sometimes one may see torque given units that don't dimensionally make sense. For example: g x cm . In these units, g should be understood as the force given by the weight of 1 gram at the surface of the earth. The surface of the earth is understood to have an average acceleration of gravity (approx. 9.80665 m/sec^2).

Special Cases and Other Facts

Moment arm Formula : A very useful special case, often given as the definition of torque in fields other than physics, is as follows:

$$|\tau| = (\text{moment arm})$$

The construction of the "moment arm" is shown right, along with the vectors r and F mentioned above. The problem with this definition is that it does not give the direction of the torque but only the magnitude, and hence it is difficult to use in three-dimensional cases. If the force is perpendicular to the displacement vector r, the moment arm will be equal to the distance to the centre, and torque will be a maximum for the given force. The equation for the magnitude of a torque, arising from a perpendicular force:

$$|\tau| = \text{(distance to centre)(force)}.$$

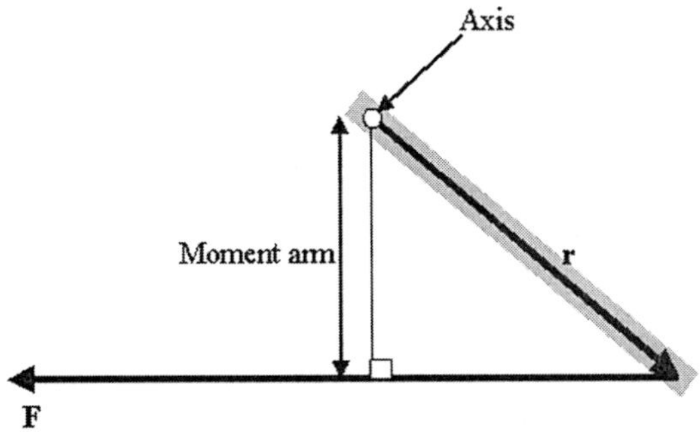

Figure 5: Moment arm diagram

For example, if a person places a force of 10 N on a spanner (wrench) which is 0.5 m long, the torque will be 5 N m, assuming that the person pulls the spanner by applying force perpendicular to the spanner.

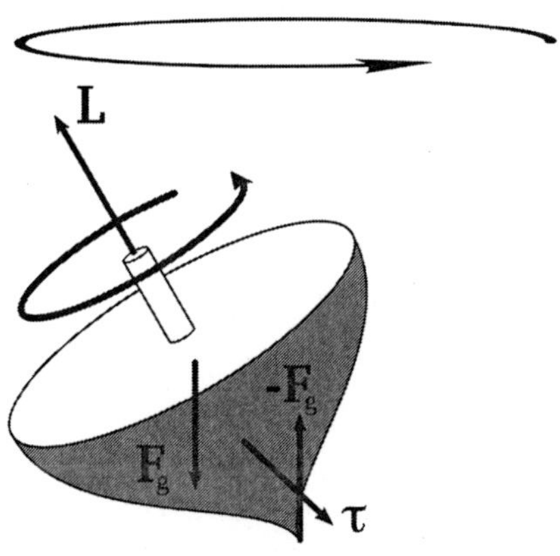

Figure 6: The torque caused by the two opposing forces F_g and $\hat{}F_g$ causes a change in the angular momentum L in the direction of that torque. This causes the top to precess.

Static Equilibrium

For an object to be in static equilibrium, not only must the sum of the forces be zero, but also the sum of the torques (moments) about any point. For a two-dimensional situation with horizontal and vertical forces, the sum of the forces requirement is two equations: $\Sigma H = 0$ and $\Sigma V = 0$, and the torque a third equation: $\Sigma\tau = 0$. That is, to solve statically determinate equilibrium problems in two-dimensions, we use three equations.

Net Force versus Torque

When the net force on the system is zero, the torque measured from any point in space is the same. For example, the torque on a current-carrying loop in a uniform magnetic field is the same regardless of your point of reference. If the net force F is not zero, and τ_1 is the torque measured from r_1, then the torque measured from r_2 is $\tau_2 = \tau_1 + (r_1 - r_2) \times F$

Machine Torque

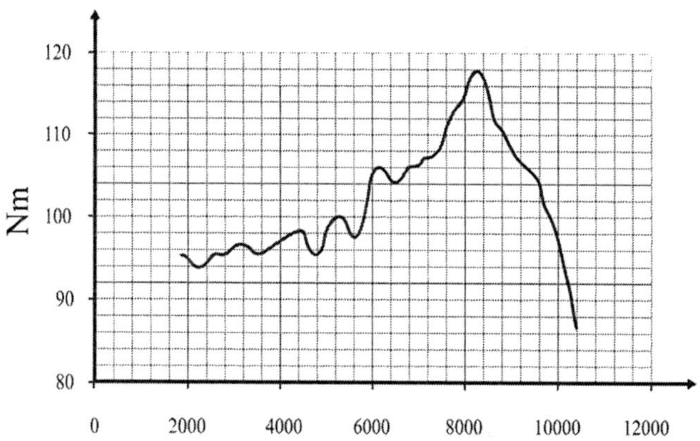

Figure 7: Torque curve of a motorcycle ("BMW K 1200 R 2005"). The horizontal axis is the speed (in rpm) that the crankshaft is turning, and the vertical axis is the torque (in Newton metres) that the engine is capable of providing at that speed.

Torque is part of the basic specification of an engine: the power output of an engine is expressed as its torque multiplied by its rotational speed of the axis. Internal-combustion engines produce useful torque only over a limited range of rotational speeds (typically from around 1,000–6,000 rpm for a small car). The varying torque output over that range can be measured with a dynamometre, and shown as a torque curve. The peak of that torque curve occurs somewhat below the overall power peak. The torque peak cannot, by definition, appear at higher rpm than the power peak.

Understanding the relationship between torque, power and engine speed is vital in automotive engineering, concerned as it is with transmitting power from the engine through the drive train to the wheels. Power is a function of torque and engine speed. The gearing of the drive train must be chosen appropriately to make the most of the motor's torque characteristics. Power at the drive wheels is equal to engine power less mechanical losses regardless of any gearing between the engine and drive wheels.

Steam engines and electric motors tend to produce maximum torque close to zero rpm, with the torque diminishing as rotational speed rises (due to increasing friction and other constraints). Reciprocating steam engines can start heavy loads from zero RPM without a clutch.

Relationship between Torque, Power and Energy

If a force is allowed to act through a distance, it is doing mechanical work. Similarly, if torque is allowed to act through a rotational distance, it is doing work. Mathematically, for rotation about a fixed axis through the centre of mass,

$$W = \int_{\theta_1}^{\theta_2} \tau \, d\theta,$$

where W is work, τ is torque, and θ_1 and θ_2 represent (respectively) the initial and final angular positions of the body. It follows from the work-energy theorem that W also represents the change in the rotational kinetic energy K_{rot} of the body, given by

$$K_{rot} = \tfrac{1}{2} I\omega^2,$$

where I is the moment of inertia of the body and ω is its angular speed.

Power is the work per unit time, given by

$$P = \tau \cdot \omega,$$

where P is power, τ is torque, ω is the angular velocity, and $\cdot$ represents the scalar product.

Mathematically, the equation may be rearranged to compute torque for a given power output. Note that the power injected by the torque depends only on the instantaneous angular speed – not on whether the angular speed increases, decreases, or remains constant while the torque is being applied (this is equivalent to the linear case where the power injected by a force depends only on the instantaneous speed – not on the resulting acceleration, if any).

In practice, this relationship can be observed in power stations which are connected to a large electrical power grid. In such an arrangement, the generator's angular speed is fixed by the grid's frequency, and the power output of the plant is determined by the torque applied to the generator's axis of rotation.

Consistent units must be used. For metric SI units power is watts, torque is newton metres and angular speed is radians per second (not rpm and not revolutions per second).

Also, the unit newton metre is dimensionally equivalent to the joule, which is the unit of energy. However, in the case of torque, the unit is assigned to a vector, whereas for energy, it is assigned to a scalar.

Conversion to Other Uunits

A conversion factor may be necessary when using different units of power, torque, or angular speed. For example, if rotational speed (revolutions per time) is used in place of angular speed (radians per time), we multiply by a factor of 2π radians per revolution.

$$\text{power} = \text{torque} \times 2\pi \times \text{rotational speed}$$

Adding units:

$$\text{power (W)} = \text{torque (N·m)} \times 2\pi \times \text{rotational speed (rps)}$$

Dividing on the left by 60 seconds per minute and by 1000 watts per kilowatt gives us the following.

$$\text{power (kW)} = \frac{\text{torque (N·m)} \times 2\pi \times \text{rotational speed (rpm)}}{60000}$$

where rotational speed is in revolutions per minute (rpm).

Some people (e.g. American automotive engineers) use horsepower (imperial mechanical) for power, foot-pounds (lbf·ft) for torque and rpm for rotational speed. This results in the formula changing to:

$$\text{power (hp)} = \frac{\text{torque(lbf·ft)} \times 2\pi \times \text{rotational speed (rpm)}}{33000}.$$

The constant below in, ft·lbf/min, changes with the definition of the horsepower; for example, using metric horsepower, it becomes ~32,550.

Use of other units (e.g. BTU/h for power) would require a different custom conversion factor.

Derivation

For a rotating object, the *linear distance* covered at the circumference in a radian of rotation is the product of the radius with the angular speed. That is: linear speed = radius × angular speed. By definition, linear distance=linear speed × time=radius × angular speed × time.

By the definition of torque: torque=force × radius. We can rearrange this to determine force=torque ÷ radius. These two values can be substituted into the definition of power:

$$\text{power} = \frac{\text{force} \times \text{linear distance}}{\text{time}} = \frac{\left(\frac{\text{torque}}{\text{r}}\right) \times (\text{r} \times \text{angular speed} \times \text{t})}{\text{t}} = \text{torque} \times$$

The radius r and time t have dropped out of the equation. However angular speed must be in radians, by the assumed direct relationship between linear speed and angular speed at the beginning of the derivation. If the rotational speed is measured in revolutions per unit of time, the linear speed and distance are increased proportionately by 2π in the above derivation to give:

$$\text{power} = \text{torque} \times 2\pi \times \text{rotational speed}.$$

If torque is in lbf·ft and rotational speed in revolutions per minute, the above equation gives power in ft·lbf/min. The horsepower form of the equation is then derived by applying the conversion factor 33000 ft·lbf/min per horsepower:

$$\text{power} = \text{torque} \times 2\pi \times \text{rotational speed} \cdot \frac{\text{ft} \cdot \text{lbf}}{\text{min}} \times \frac{\text{horsepower}}{33000 \cdot \frac{\text{ft} \cdot \text{lbf}}{\text{min}}} \approx \frac{\text{torque} \times \text{RPM}}{5252}$$

$$\text{because} \quad 5252.113122... = \frac{33000}{2\pi}.$$

Principle of Moments

The Principle of Moments, also known as Varignon's theorem (not to be confused with the geometrical theorem of the same name) states that the sum of torques due to several forces applied to *a single* point is equal to the torque due to the sum (resultant) of the forces. Mathematically, this follows from:

$$(\text{r} \times \text{F}_1) + (\text{r} \times \text{F}_2) + \cdots = \text{r} \times (\text{F}_1 + \text{F}_2 + \cdots).$$

Torque Multiplier

A torque multiplier is a gear box with reduction ratios greater than 1. The given torque at the input gets multiplied as per the reduction ratio and transmitted to the output, thereby achieving greater torque, but with reduced rotational speed.

Chapter 7

Fluid Theory

Fluid Dynamics

Fluid dynamics is a sub-discipline of fluid mechanics that deals with fluid flow—the natural science of fluids (liquids and gases) in motion. It has several subdisciplines itself, including aerodynamics (the study of air and other gases in motion) and hydrodynamics (the study of liquids in motion). Fluid dynamics has a wide range of applications, including calculating forces and moments on aircraft, determining the mass flow rate of petroleum through pipelines, predicting weather patterns, understanding nebulae in interstellar space and reportedly modelling fission weapon detonation. Some of its principles are even used in traffic engineering, where traffic is treated as a continuous fluid.

Fluid dynamics offers a systematic structure that underlies these practical disciplines, that embraces empirical and semi-empirical laws derived from flow measurement and used to solve practical problems. The solution to a fluid dynamics problem typically involves calculating various properties of the fluid, such as velocity, pressure, density, and temperature, as functions of space and time.

Historically, *hydrodynamics* meant something different than it does today. Before the twentieth century, hydrodynamics was synonymous with fluid dynamics. This is still reflected in names of some fluid dynamics topics, like magnetohydrodynamics and hydrodynamic stability—both also applicable in, as well as being applied to, gases.

Equations of Fluid Dynamics

The foundational axioms of fluid dynamics are the conservation laws, specifically, conservation of mass, conservation of linear momentum (also known as Newton's Second Law of Motion), and conservation of energy (also known as First Law of Thermodynamics).

These are based on classical mechanics and are modified in quantum mechanics and general relativity. They are expressed using the Reynolds Transport Theorem.

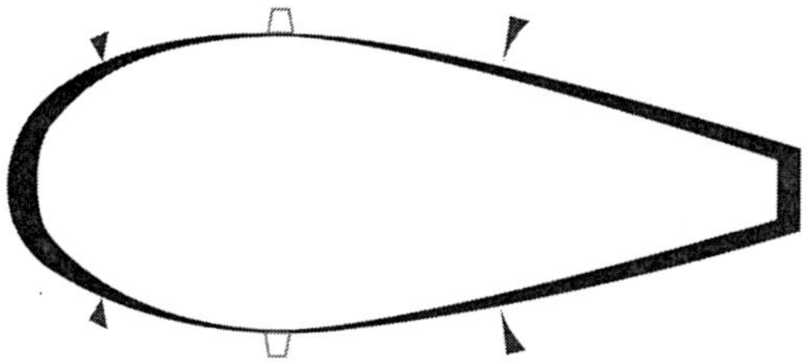

Figure 1: Typical aerodynamic teardrop shape, assuming a viscous medium passing from left to right, the diagram shows the pressure distribution as the thickness of the black line and shows the velocity in the boundary layer as the violet triangles. The green vortex generators prompt the transition to turbulent flow and prevent back-flow also called flow separation from the high pressure region in the back. The surface in front is as smooth as possible or even employs shark like skin, as any turbulence here will reduce the energy of the airflow. The truncation on the right, known as a Kamm back, also prevents back flow from the high pressure region in the back across the spoilers to the convergent part.

In addition to the above, fluids are assumed to obey the *continuum assumption*. Fluids are composed of molecules that collide with one another and solid objects. However, the continuum assumption considers fluids to be continuous, rather than discrete. Consequently, properties such as density, pressure, temperature, and velocity are taken to be well-defined at infinitesimally small points, and are assumed to vary continuously from one point to another. The fact that the fluid is made up of discrete molecules is ignored.

For fluids which are sufficiently dense to be a continuum, do not contain ionised species, and have velocities small in relation to the speed of light, the momentum equations for Newtonian fluids are the Navier-Stokes equations, which is a non-linear set of differential equations that describes the flow of a fluid whose stress depends linearly on velocity gradients and pressure. The unsimplified equations do not have a general closed-form solution, so they are primarily of use in Computational Fluid Dynamics. The equations can be simplified in a number of ways, all of which make them easier to solve. Some of them allow appropriate fluid dynamics problems to be solved in closed form.

In addition to the mass, momentum, and energy conservation equations, a thermodynamical equation of state giving the pressure as a function of other thermodynamic variables for the fluid is required to completely specify the problem. An example of this would be the perfect gas equation of state:

$$p = \frac{\rho R_u T}{M}$$

where p is pressure, ρ is density, R_u is the gas constant, M is the molar mass and T is temperature.

Compressible vs Incompressible Flow

All fluids are compressible to some extent, that is changes in pressure or temperature will result in changes in density. However, in many situations the changes in pressure and temperature are sufficiently small that the changes in density are negligible. In this case the flow can be modelled as an incompressible flow. Otherwise the more general compressible flow equations must be used.

Mathematically, incompressibility is expressed by saying that the density ρ of a fluid parcel does not change as it moves in the flow field, i.e.,

$$\frac{D\rho}{Dt} = 0,$$

where D / Dt is the substantial derivative, which is the sum of local and convective derivatives. This additional constraint simplifies the governing equations, especially in the case when the fluid has a uniform density.

For flow of gases, to determine whether to use compressible or incompressible fluid dynamics, the Mach number of the flow is to be evaluated. As a rough guide, compressible effects can be ignored at Mach numbers below approximately 0.3. For liquids, whether the incompressible assumption is valid depends on the fluid properties (specifically the critical pressure and temperature of the fluid) and the flow conditions (how close to the critical pressure the actual flow pressure becomes). Acoustic problems always require allowing compressibility, since sound waves are compression waves involving changes in pressure and density of the medium through which they propagate.

Viscous vs Inviscid Flow

Viscous problems are those in which fluid friction has significant effects on the fluid motion.

The Reynolds number, which is a ratio between inertial and viscous forces, can be used to evaluate whether viscous or inviscid equations are appropriate to the problem.

Stokes flow is flow at very low Reynolds numbers, *Re*<<1, such that inertial forces can be neglected compared to viscous forces.

On the contrary, high Reynolds numbers indicate that the inertial forces are more significant than the viscous (friction) forces. Therefore, we may assume the flow to be an inviscid flow, an approximation in which we neglect viscosity completely, compared to inertial terms.

This idea can work fairly well when the Reynolds number is high. However, certain problems such as those involving solid boundaries, may require that the viscosity be included. Viscosity often cannot be neglected near solid boundaries because the no-slip condition can generate a thin region of large strain rate (known as Boundary layer) which enhances the effect of even a small amount of viscosity, and thus generating vorticity. Therefore, to calculate net forces on bodies (such as wings) we should use viscous flow equations. As illustrated by d'Alembert's paradox, a body in an inviscid fluid will experience no drag force. The standard equations of inviscid flow are the Euler equations. Another often used model, especially in computational fluid dynamics, is to use the Euler equations away from the body and the boundary layer equations, which incorporates viscosity, in a region close to the body.

The Euler equations can be integrated along a streamline to get Bernoulli's equation. When the flow is everywhere irrotational and inviscid, Bernoulli's equation can be used throughout the flow field. Such flows are called potential flows.

Steady vs Unsteady Flow

When all the time derivatives of a flow field vanish, the flow is considered to be a steady flow. Steady-state flow refers to the condition where the fluid properties at a point in the system do not change over time. Otherwise, flow is called unsteady. Whether a particular flow is steady or unsteady, can depend on the chosen frame of reference. For instance, laminar flow over a sphere is steady in the frame of reference that is stationary with respect to the sphere. In a frame of reference that is stationary with respect to a background flow, the flow is unsteady.

Turbulent flows are unsteady by definition. A turbulent flow can, however, be statistically stationary. According to Pope:

The random field U(x,t) *is statistically stationary if all statistics are invariant under a shift in time.*

This roughly means that all statistical properties are constant in time. Often, the mean field is the object of interest, and this is constant too in a statistically stationary flow.

Steady flows are often more tractable than otherwise similar unsteady flows. The governing equations of a steady problem have one dimension fewer (time) than the governing equations of the same problem without taking advantage of the steadiness of the flow field.

Laminar vs Turbulent Flow

Turbulence is flow characterised by recirculation, eddies, and apparent randomness. Flow in which turbulence is not exhibited is called laminar. It should be noted, however, that the presence of eddies or recirculation alone does not necessarily indicate turbulent flow—these phenomena may be present in laminar flow as well. Mathematically, turbulent flow is often represented via a Reynolds decomposition, in which the flow is broken down into the sum of an average component and a perturbation component.

It is believed that turbulent flows can be described well through the use of the Navier–Stokes equations. Direct numerical simulation (DNS), based on the Navier–Stokes equations, makes it possible to simulate turbulent flows at moderate Reynolds numbers. Restrictions depend on the power of the computer used and the efficiency of the solution algorithm. The results of DNS have been found to agree well with experimental data for some flows.

Most flows of interest have Reynolds numbers much too high for DNS to be a viable option, given the state of computational power for the next few decades. Any flight vehicle large enough to carry a human (L > 3 m), moving faster than 72 km/h (20 m/s) is well beyond the limit of DNS simulation (Re = 4 million). Transport aircraft wings (such as on an Airbus A300 or Boeing 747) have Reynolds numbers of 40 million (based on the wing chord). In order to solve these real-life flow problems, turbulence models will be a necessity for the foreseeable future. Reynolds-averaged Navier–Stokes equations (RANS) combined with turbulence modelling provides a model of the effects of the turbulent flow. Such a modelling mainly provides the additional momentum transfer by the Reynolds stresses, although the turbulence also enhances the heat and mass transfer. Another promising methodology is large eddy simulation (LES), especially in the guise of detached eddy simulation (DES)—which is a combination of RANS turbulence modelling and large eddy simulation.

Newtonian vs non-Newtonian Fluids

Sir Isaac Newton showed how stress and the rate of strain are very close to linearly related for many familiar fluids, such as water and air. These Newtonian fluids are modelled by a coefficient called viscosity, which depends on the specific fluid.

However, some of the other materials, such as emulsions and slurries and some visco-elastic materials (e.g. blood, some polymers), have more complicated *non-Newtonian* stress-strain behaviours. These materials include *sticky liquids* such as latex, honey, and lubricants which are studied in the sub-discipline of rheology.

Subsonic vs Transonic, Supersonic and Hypersonic Flows

While many terrestrial flows (e.g. flow of water through a pipe) occur at low mach numbers, many flows of practical interest (e.g. in aerodynamics) occur at high fractions of the Mach Number M=1 or in excess of it (supersonic flows). New phenomena occur at these Mach number regimes (e.g. shock waves for supersonic flow, transonic instability in a regime of flows with M nearly equal to 1, non-equilibrium chemical behaviour due to ionisation in hypersonic flows) and it is necessary to treat each of these flow regimes separately.

Magnetohydrodynamics

Magnetohydrodynamics is the multi-disciplinary study of the flow of electrically conducting fluids in electromagnetic fields. Examples of such fluids include plasmas, liquid metals, and salt water. The fluid flow equations are solved simultaneously with Maxwell's equations of electromagnetism.

Other Approximations

There are a large number of other possible approximations to fluid dynamic problems. Some of the more commonly used are listed below.

- The Boussinesq approximation neglects variations in density except to calculate buoyancy forces. It is often used in free convection problems where density changes are small.
- Lubrication theory and Hele-Shaw flow exploits the large aspect ratio of the domain to show that certain terms in the equations are small and so can be neglected.

- Slender-body theory is a methodology used in Stokes flow problems to estimate the force on, or flow field around, a long slender object in a viscous fluid.
- The shallow-water equations can be used to describe a layer of relatively inviscid fluid with a free surface, in which surface gradients are small.
- The Boussinesq equations are applicable to surface waves on thicker layers of fluid and with steeper surface slopes.
- Darcy's law is used for flow in porous media, and works with variables averaged over several pore-widths.
- In rotating systems, the quasi-geostrophic approximation assumes an almost perfect balance between pressure gradients and the Coriolis force. It is useful in the study of atmospheric dynamics.

Terminology in Fluid Dynamics

The concept of pressure is central to the study of both fluid statics and fluid dynamics. A pressure can be identified for every point in a body of fluid, regardless of whether the fluid is in motion or not. Pressure can be measured using an aneroid, Bourdon tube, mercury column, or various other methods.

Some of the terminology that is necessary in the study of fluid dynamics is not found in other similar areas of study. In particular, some of the terminology used in fluid dynamics is not used in fluid statics.

Terminology in Incompressible Fluid Dynamics

The concepts of total pressure and dynamic pressure arise from Bernoulli's equation and are significant in the study of all fluid flows. (These two pressures are not pressures in the usual sense—they cannot be measured using an aneroid, Bourdon tube or mercury column.) To avoid potential ambiguity when referring to pressure in fluid dynamics, many authors use the term static pressure to distinguish it from total pressure and dynamic pressure. Static pressure is identical to pressure and can be identified for every point in a fluid flow field.

In Aerodynamics, L.J. Clancy writes: To distinguish it from the total and dynamic pressures, the actual pressure of the fluid, which is associated not with its motion but with its state, is often referred to as the static pressure, but where the term pressure alone is used it refers to this static pressure.

A point in a fluid flow where the flow has come to rest (i.e. speed is equal to zero adjacent to some solid body immersed in the fluid flow) is of special significance. It is of such importance that it is given a special name—a stagnation point. The static pressure at the stagnation point is of special significance and is given its own name—stagnation pressure. In incompressible flows, the stagnation pressure at a stagnation point is equal to the total pressure throughout the flow field.

Terminology in Compressible Fluid Dynamics

In a compressible fluid, such as air, the temperature and density are essential when determining the state of the fluid. In addition to the concept of total pressure (also known as stagnation pressure), the concepts of total (or stagnation) temperature and total (or stagnation) density are also essential in any study of compressible fluid flows. To avoid potential ambiguity when referring to temperature and density, many authors use the terms static temperature and static density. Static temperature is identical to temperature; and static density is identical to density; and both can be identified for every point in a fluid flow field. The temperature and density at a stagnation point are called stagnation temperature and stagnation density.

A similar approach is also taken with the thermodynamic properties of compressible fluids. Many authors use the terms total (or stagnation) enthalpy and total (or stagnation) entropy. The terms static enthalpy and static entropy appear to be less common, but where they are used they mean nothing more than enthalpy and entropy respectively, and the prefix "static" is being used to avoid ambiguity with their 'total' or 'stagnation' counterparts. Because the 'total' flow conditions are defined by is entropically bringing the fluid to rest, the total (or stagnation) entropy is by definition always equal to the "static" entropy.

Fluid Statics

Fluid statics (also called hydrostatics) is the science of fluids at rest, and is a sub-field within fluid mechanics. The term usually refers to the mathematical treatment of the subject. It embraces the study of the conditions under which fluids are at rest in stable equilibrium. The use of fluid to do work is called hydraulics, and the science of fluids in motion is fluid dynamics.

Pressure in Fluids at Rest

Due to the fundamental nature of fluids, a fluid cannot remain at rest under the presence of a shear stress. However, fluids can exert pressure normal to any contacting surface. If a point in the fluid is thought of as an infinitesimally small cube, then it follows from the principles of equilibrium that the pressure on every side of this unit of fluid must be equal. If this were not the case, the fluid would move in the direction of the resulting force. Thus, the pressure on a fluid at rest is isotropic; i.e., it acts with equal magnitude in all directions. This characteristic allows fluids to transmit force through the length of pipes or tubes; i.e., a force applied to a fluid in a pipe is transmitted, via the fluid, to the other end of the pipe.

This concept was first formulated, in a slightly extended form, by the French mathematician and philosopher Blaise Pascal in 1647 and would later be known as Pascal's law. This law has many important applications in hydraulics.

Hydrostatic Pressure

Hydrostatic pressure is the pressure exerted by a fluid at equilibrium due to the force of gravity. A fluid in this condition is known as a hydrostatic fluid. The hydrostatic pressure can be determined from a control volume analysis of an infinitesimally small cube of fluid. Since pressure is defined as the force exerted on a test area ($p = F/A$, with p: pressure, F: force normal to area A, A: area), and the only force acting on any such small cube of fluid is the weight of the fluid column above it, hydrostatic pressure can be calculated according to the following formula:

$$p(z) = \frac{1}{A}\int_{z_0}^{z} dz' \iint_A dx'dy' \rho(z')g(z') = \int_{z_0}^{z} dz' \rho(z')g(z'),$$

where:

- p is the hydrostatic pressure (Pa),
- ρ is the fluid density (kg/m^3),
- g is gravitational acceleration (m/s^2),
- A is the test area (m^2),
- z is the height (parallel to the direction of gravity) of the test area (m),
- z_0 is the height of the zero reference point of the pressure (m).

For water and other liquids, this integral can be simplified significantly for many practical applications, based on the following two assumptions: Since many liquids can be considered incompressible, a reasonably good estimation can be made from assuming a constant density throughout the liquid. (The same assumption cannot be made within a gaseous environment.) Also, since the height h of the fluid column between z and z_0 is often reasonably small compared to the radius of the Earth, one can neglect the variation of g. Under these circumstances, the integral boils down to the simple formula:

$$p = \rho g h,$$

where h is the height z-z_0 of the liquid column between the test volume and the zero reference point of the pressure. Note that this reference point should lie at or below the surface of the liquid. Otherwise, one has to split the integral into two (or more) terms with the constant ρ_{liquid} and $\rho(z')_{above}$. For example, the absolute pressure compared to vacuum is :

$$p = \rho g H + p_{atm}$$

where H is the total height of the liquid column above the test area the surface, and p_{atm} is the atmospheric pressure, i.e., the pressure calculated from the remaining integral over the air column from the liquid surface to infinity.

Hydrostatic pressure has been used in the preservation of foods in a process called pascalisation.

Atmospheric Pressure

Statistical mechanics shows that, for a gas of constant temperature, T, its pressure, p will vary with height, h, as:

$$p(h) = p(0)e^{-Mgh/kT}$$

where:

g = the acceleration due to gravity

T = Absolute temperature

k = Boltzmann constant

M = mass of a single molecule of gas

p = pressure

h = height

This is known as the barometric formula, and may be derived from assuming the pressure is hydrostatic.

If there are multiple types of molecules in the gas, the partial pressure of each type will be given by this equation. Under most conditions, the distribution of each species of gas is independent of the other species.

Buoyancy

Any body of arbitrary shape which is immersed, partly or fully, in a fluid will experience the action of a net force in the opposite direction of the local pressure gradient. If this pressure gradient arises from gravity, the net force is in the vertical direction opposite that of the gravitational force. This vertical force is termed buoyancy or buoyant force and is equal in magnitude, but opposite in direction, to the weight of the displaced fluid.

In the case of a ship, for instance, its weight is balanced by shear force from the displaced water, allowing it to float. If more cargo is loaded onto the ship, it would sink more into the water - displacing more water and thus receive a higher buoyant force to balance the increased weight.

Discovery of the principle of buoyancy is attributed to Archimedes.

Liquids-fluids with Free Surfaces

Liquids can have free surfaces at which they interface with gases, or with a vacuum. In general, the lack of the ability to sustain a shear stress entails that free surfaces rapidly adjust towards an equilibrium. However, on small length scales, there is an important balancing force from surface tension.

Capillary Action

When liquids are constrained in vessels whose dimensions are small, compared to the relevant length scales, surface tension effects become important leading to the formation of a meniscus through capillary action. This capillary action has profound consequences for biological systems as it is part of one of the two driving mechanisms of the flow of water in plant xylem, the transpirational pull.

Drops

Without surface tension, drops would not be able to form. The dimensions and stability of drops are determined by surface tension. The drop's surface tension is directly proportional to the cohesion property of the fluid.

Surface Tension

Surface tension is a property of the surface of a liquid that allows it to resist an external force. It is revealed, for example, in floating of some objects on the surface of water, even though they are denser than water, and in the ability of some insects (e.g. water striders) to run on the water surface. This property is caused by cohesion of like molecules, and is responsible for many of the behaviours of liquids.

Surface tension has the dimension of force per unitlength, or of energy per unit area. The two are equivalent—but when referring to energy per unit of area, people use the term surface energy—which is a more general term in the sense that it applies also to solids and not just liquids.

In materials science, surface tension is used for either surface stress or surface free energy.

Cause

The cohesive forces among the liquid molecules are responsible for this phenomenon of surface tension. In the bulk of the liquid, each molecule is pulled equally in every direction by neighbouring liquid molecules, resulting in a net force of zero. The molecules at the surface do not have other molecules on all sides of them and therefore are pulled inwards. This creates some internal pressure and forces liquid surfaces to contract to the minimal area.

Surface tension is responsible for the shape of liquid droplets. Although easily deformed, droplets of water tend to be pulled into a spherical shape by the cohesive forces of the surface layer. In the absence of other forces, including gravity, drops of virtually all liquids would be perfectly spherical. The spherical shape minimises the necessary "wall tension" of the surface layer according to Laplace's law.

Another way to view it is in terms of energy. A molecule in contact with a neighbour is in a lower state of energy than if it were alone (not in contact with a neighbour). The interior molecules have as many neighbours as they can possibly have, but the boundary molecules are missing neighbours (compared to interior molecules) and therefore have a higher energy. For the liquid to minimise its energy state, the number of higher energy boundary molecules must be minimised. The minimised quantity of boundary molecules results in a minimised surface area.

As a result of surface area minimisation, a surface will assume the smoothest shape it can (mathematical proof that "smooth" shapes minimise surface area relies on use of the Euler–Lagrange equation). Since any curvature in the surface shape results in greater area, a higher energy will also result. Consequently the surface will push back against any curvature in much the same way as a ball pushed uphill will push back to minimise its gravitational potential energy.

Effects in Everyday Life

Water

Several effects of surface tension can be seen with ordinary water:

A. Beading of rain water on the surface of a waxy surface, such as a leaf. Water adheres weakly to wax and strongly to itself, so water clusters into drops. Surface tension gives them their near-spherical shape, because a sphere has the smallest possible surface area to volume ratio.

B. Formation of drops occurs when a mass of liquid is stretched. The animation shows water adhering to the faucet gaining mass until it is stretched to a point where the surface tension can no longer bind it to the faucet. It then separates and surface tension forms the drop into a sphere. If a stream of water were running from the faucet, the stream would break up into drops during its fall. Gravity stretches the stream, then surface tension pinches it into spheres.

C. Floatation of objects denser than water occurs when the object is nonwettable and its weight is small enough to be borne by the forces arising from surface tension. For example, water striders use surface tension to walk on the surface of a pond. The surface of the water behaves like an elastic film: the insect's feet cause indentations in the water's surface, increasing its surface area.

D. Separation of oil and water (in this case, water and liquid wax) is caused by a tension in the surface between dissimilar liquids. This type of surface tension is called "interface tension", but its physics are the same.

E. Tears of wine is the formation of drops and rivulets on the side of a glass containing an alcoholic beverage. Its cause is a complex interaction between the differing surface tensions of water and ethanol; it is induced by a combination of surface tension modification of water by ethanol together with ethanol evaporating faster than water.

Surfactants

Surface tension is visible in other common phenomena, especially when surfactants are used to decrease it:

- Soap bubbles have very large surface areas with very little mass. Bubbles in pure water are unstable. The addition of surfactants, however, can have a stabilising effect on the bubbles. Notice that surfactants actually reduce the surface tension of water by a factor of three or more.

- Emulsions are a type of solution in which surface tension plays a role. Tiny fragments of oil suspended in pure water will spontaneously assemble themselves into much larger masses. But the presence of a surfactant provides a decrease in surface tension, which permits stability of minute droplets of oil in the bulk of water (or vice versa).

Basic Physics

Surface tension, represented by the symbol γ is defined as the force along a line of unit length, where the force is parallel to the surface but perpendicular to the line. One way to picture this is to imagine a flat soap film bounded on one side by a taut thread of length, L. The thread will be pulled toward the interior of the film by a force equal to $2\gamma L$ (the factor of 2 is because the soap film has two sides, hence two surfaces). Surface tension is therefore measured in forces per unit length. Its SI unit is newton per metre but the cgs unit of dyne per cm is also used. One dyn/cm corresponds to 0.001 N/m.

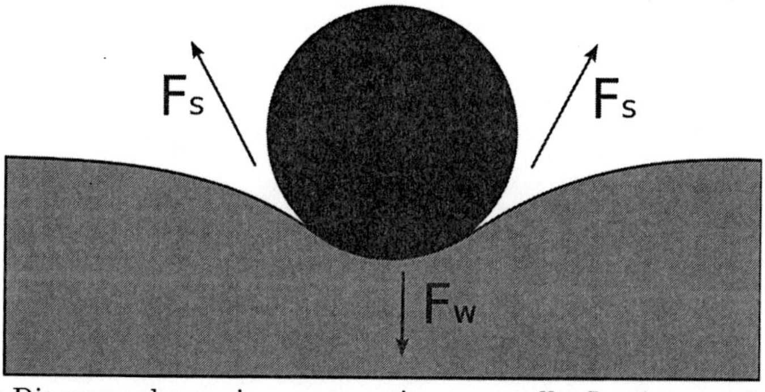

Figure 2: Diagram shows, in cross-section, a needle floating on the surface of water. Its weight, F_w, depresses the surface, and is balanced by the surface tension forces on either side, F_s, which are each parallel to the water's surface at the points where it contacts the needle. Notice that the horizontal components of the two F_s arrows point in opposite directions, so they cancel each other, but the vertical components point in the same direction and therefore add up to balance F_w.

An equivalent definition, one that is useful in thermodynamics, is work done per unit area. As such, in order to increase the surface area of a mass of liquid by an amount, δA, a quantity of work, δA, is needed. This work is stored as potential energy. Consequently surface tension can be also measured in SI system as joules per square metre and in the cgs system as ergs per cm^2. Since mechanical systems try to find a state of minimum potential energy, a free droplet of liquid naturally assumes a spherical shape, which has the minimum surface area for a given volume.

The equivalence of measurement of energy per unit area to force per unit length can be proven by dimensional analysis.

Surface Survature and Pressure

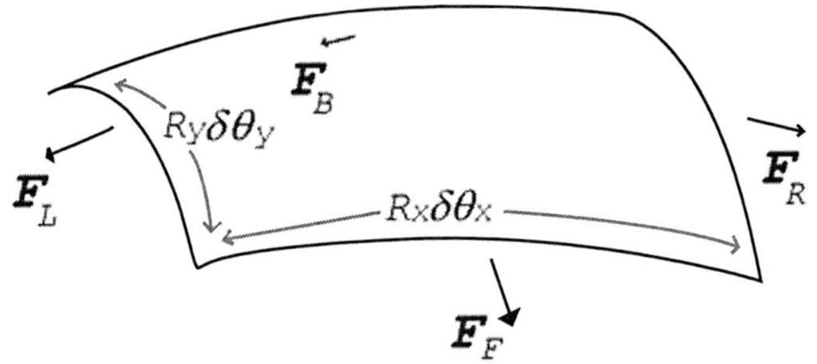

Figure 3: Surface tension forces acting on a tiny (differential) patch of surface. $\delta\theta_x$ and $\delta\theta_y$ indicate the amount of bend over the dimensions of the patch. Balancing the tension forces with pressure leads to the Young–Laplace equation

If no force acts normal to a tensioned surface, the surface must remain flat. But if the pressure on one side of the surface differs from pressure on the other side, the pressure difference times surface area results in a normal force. In order for the surface tension forces to cancel the force due to pressure, the surface must be curved. The diagram shows how surface curvature of a tiny patch of surface leads to a net component of surface tension forces acting normal to the centre of the patch. When all the forces are balanced, the resulting equation is known as the Young–Laplace equation:

$$\Delta p = \gamma\left(\frac{1}{R_x} + \frac{1}{R_y}\right)$$

where:

- Δp is the pressure difference.
- is surface tension.
- R_x and R_y are radii of curvature in each of the axes that are parallel to the surface.

The quantity in parentheses on the right hand side is in fact (twice) the mean curvature of the surface (depending on normalisation).

Solutions to this equation determine the shape of water drops, puddles, menisci, soap bubbles, and all other shapes determined by surface tension (such as the shape of the impressions that a water strider's feet make on the surface of a pond).

The below shows how the internal pressure of a water droplet increases with decreasing radius. For not very small drops the effect is subtle, but the pressure difference becomes enormous when the drop sizes approach the molecular size. (In the limit of a single molecule the concept becomes meaningless.)

Table 1: D*p* for water drops of different radii at STP

Droplet radius	1 mm	0.1 mm	1 μm	10 nm
Δp (atm)	0.0014	0.0144	1.436	143.6

Liquid Surface

Figure 4: Minimal surface

To find the shape of the minimal surface bounded by some arbitrary shaped frame using strictly mathematical means can be a daunting task. Yet by fashioning the frame out of wire and dipping it in soap-solution, a locally minimal surface will appear in the resulting soap-film within seconds.

The reason for this is that the pressure difference across a fluid interface is proportional to the mean curvature, as seen in the Young-Laplace equation. For an open soap film, the pressure difference is zero, hence the mean curvature is zero, and minimal surfaces have the property of zero mean curvature.

Contact Angles

The surface of any liquid is an interface between that liquid and some other medium. The top surface of a pond, for example, is an interface between the pond water and the air. Surface tension, then, is not a property of the liquid alone, but a property of the liquid's interface with another medium. If a liquid is in a container, then besides the liquid/air interface at its top surface, there is also an interface between the liquid and the walls of the container. The surface tension between the liquid and air is usually different (greater than) its surface tension with the walls of a container. And where the two surfaces meet, their geometry must be such that all forces balance.

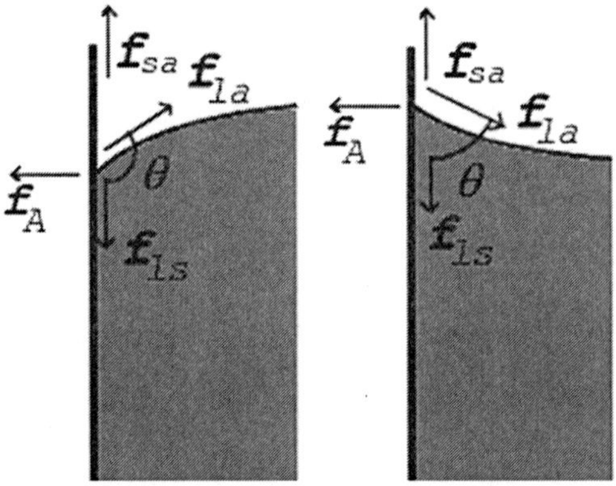

Figure 5: Forces at contact point shown for contact angle greater than 90° (left) and less than 90° (right)

Where the two surfaces meet, they form a contact angle, θ, which is the angle the tangent to the surface makes with the solid surface. The diagram to the right shows two examples. Tension forces are shown for the liquid-air interface, the liquid-solid interface, and the solid-air interface. The example on the left is where the difference between the liquid-solid and solid-air surface tension, $\gamma_{ls} - \gamma_{sa}$, is less than the liquid-air surface tension, γ_{la}, but is nevertheless positive, that is

$$\gamma_{la} > \gamma_{ls} - \gamma_{sa} > 0$$

In the diagram, both the vertical and horizontal forces must cancel exactly at the contact point. The horizontal component of f_{la} is cancelled by the adhesive force, f_A.

$$f_A = f_{la} \sin\theta$$

The more telling balance of forces, though, is in the vertical direction. The vertical component of f_{la} must exactly cancel the force, f_{ls}.

$$f_{ls} - f_{sa} = -f_{la}\cos\theta$$

Since the forces are in direct proportion to their respective surface tensions, we also have:

$$\gamma_{ls} - \gamma_{sa} = -\gamma_{la}\cos\theta$$

where

- γ_{ls} is the liquid-solid surface tension,
- γ_{la} is the liquid-air surface tension,
- γ_{sa} is the solid-air surface tension,
- θ is the contact angle, where a concavemeniscus has contact angle less than 90° and a convex meniscus has contact angle of greater than 90°.

This means that although the difference between the liquid-solid and solid-air surface tension, $\gamma_{ls} - \gamma_{sa}$, is difficult to measure directly, it can be inferred from the easily measured contact angle, θ, if the liquid-air surface tension, γ_{la}, is known.

This same relationship exists in the diagram on the right. But in this case see that because the contact angle is less than 90°, the liquid-solid/solid-air surface tension difference must be negative:

$$\gamma_{la} > 0 > \gamma_{ls} - \gamma_{sa}$$

Special Contact Angles

Observe that in the special case of a water-silver interface where the contact angle is equal to 90°, the liquid-solid/solid-air surface tension difference is exactly zero.

Another special case is where the contact angle is exactly 180°. Water with specially prepared Teflon approaches this. Contact angle of 180° occurs when the liquid-solid surface tension is exactly equal to the liquid-air surface tension.

$$\gamma_{la} = \gamma_{ls} - \gamma_{sa} > 0 \qquad \theta = 180^{\circ}$$

Methods of Measurement

Because surface tension manifests itself in various effects, it offers a number of paths to its measurement. Which method is optimal depends upon the nature of the liquid being measured, the conditions under which its tension is to be measured, and the stability of its surface when it is deformed.

- Du Noüy Ring method: The traditional method used to measure surface or interfacial tension. Wetting properties of the surface or interface have little influence on this measuring technique. Maximum pull exerted on the ring by the surface is measured.
- Du Noüy-Padday method: A minimised version of Du Noüy method uses a small diametre metal needle instead of a ring, in combination with a high sensitivity microbalance to record maximum pull. The advantage of this method is that very small sample volumes (down to few tens of microlitres) can be measured with very high precision, without the need to correct for buoyancy (for a needle or rather, rod, with proper geometry). Further, the measurement can be performed very quickly, minimally in about 20 seconds. First commercial multichannel tensiometres [CMCeeker] were recently built based on this principle.
- Wilhelmy plate method: A universal method especially suited to check surface tension over long time intervals. A vertical plate of known perimetre is attached to a balance, and the force due to wetting is measured.
- Spinning drop method: This technique is ideal for measuring low interfacial tensions. The diametre of a drop within a heavy phase is measured while both are rotated.
- Pendant drop method: Surface and interfacial tension can be measured by this technique, even at elevated temperatures and pressures. Geometry of a drop is analysed optically.
- Bubble pressure method (Jaeger's method): A measurement technique for determining surface tension at short surface ages. Maximum pressure of each bubble is measured.
- Drop volume method: A method for determining interfacial tension as a function of interface age. Liquid of one density is pumped into a second liquid of a different density and time between drops produced is measured.
- Capillary rise method: The end of a capillary is immersed into the solution. The height at which the solution reaches inside the capillary is related to the surface tension by the equation discussed below.

- Stalagmometric method: A method of weighting and reading a drop of liquid.
- Sessile drop method: A method for determining surface tension and density by placing a drop on a substrate and measuring the contact angle.
- Vibrational frequency of levitated drops: The natural frequency of vibrational oscillations of magnetically levitated drops has been used to measure the surface tension of superfluid ^{4}He. This value is estimated to be 0.375 dyn/cm at T = 0 K.

Effects

Liquid in a Vertical Tube: An old style mercury barometre consists of a vertical glass tube about 1 cm in diametre partially filled with mercury, and with a vacuum (called Torricelli's vacuum) in the unfilled volume. Notice that the mercury level at the centre of the tube is higher than at the edges, making the upper surface of the mercury dome-shaped. The centre of mass of the entire column of mercury would be slightly lower if the top surface of the mercury were flat over the entire crossection of the tube. But the dome-shaped top gives slightly less surface area to the entire mass of mercury. Again the two effects combine to minimise the total potential energy. Such a surface shape is known as a convex meniscus.

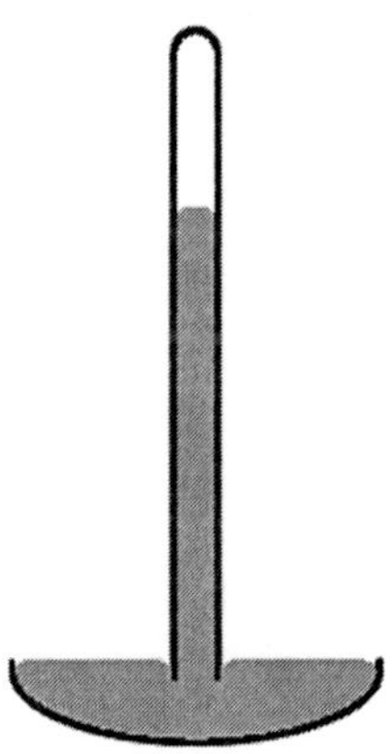

Figure 6: Diagram of a mercury barometre

The reason we consider the surface area of the entire mass of mercury, including the part of the surface that is in contact with the glass, is because mercury does not adhere at all to glass. So the surface tension of the mercury acts over its entire surface area, including where it is in contact with the glass. If instead of glass, the tube were made out of copper, the situation would be very different. Mercury aggressively adheres to copper. So in a copper tube, the level of mercury at the centre of the tube will be lower than at the edges (that is, it would be a concave meniscus). In a situation where the liquid adheres to the walls of its container, we consider the part of the fluid's surface area that is in contact with the container to have *negative* surface tension. The fluid then works to maximise the contact surface area. So in this case increasing the area in contact with the container decreases rather than increases the potential energy. That decrease is enough to compensate for the increased potential energy associated with lifting the fluid near the walls of the container.

If a tube is sufficiently narrow and the liquid adhesion to its walls is sufficiently strong, surface tension can draw liquid up the tube in a phenomenon known as capillary action. The height the column is lifted to is given by:

$$h = \frac{2\gamma_{la}\cos\theta}{\rho g r}$$

where

- h is the height the liquid is lifted,
- γ_{la} is the liquid-air surface tension,
- ρ is the density of the liquid,
- r is the radius of the capillary,
- g is the acceleration due to gravity,
- θ is the angle of contact described above. If θ is greater than 90°, as with mercury in a glass container, the liquid will be depressed rather than lifted.

Puddles on a Surface

Pouring mercury onto a horizontal flat sheet of glass results in a puddle that has a perceptible thickness. The puddle will spread out only to the point where it is a little under half a centimetre thick, and no thinner. Again this is due to the action of mercury's strong surface tension.

The liquid mass flattens out because that brings as much of the mercury to as low a level as possible, but the surface tension, at the same time, is acting to reduce the total surface area. The result is the compromise of a puddle of a nearly fixed thickness.

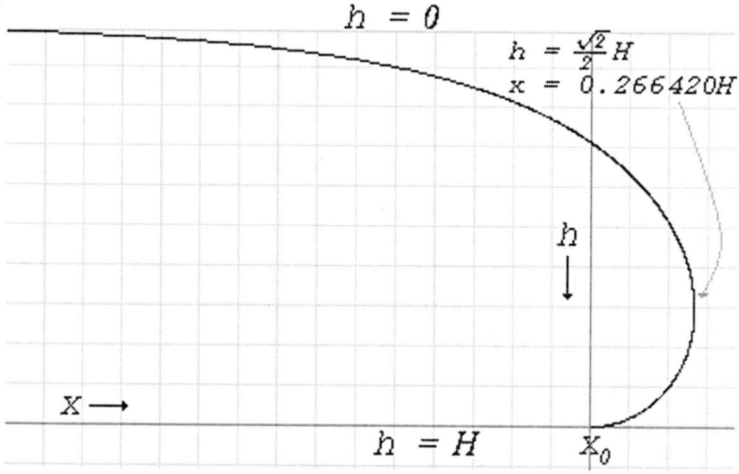

Figure 7: Profile curve of the edge of a puddle where the contact angle is 180°.

The curve is given by the formula: $x - x_0 = \frac{1}{2} H \cosh^{-1}\left(\frac{H}{h}\right) - H\sqrt{1 - \frac{h^2}{H^2}}$ where $H = 2\sqrt{\frac{\gamma}{g\rho}}$

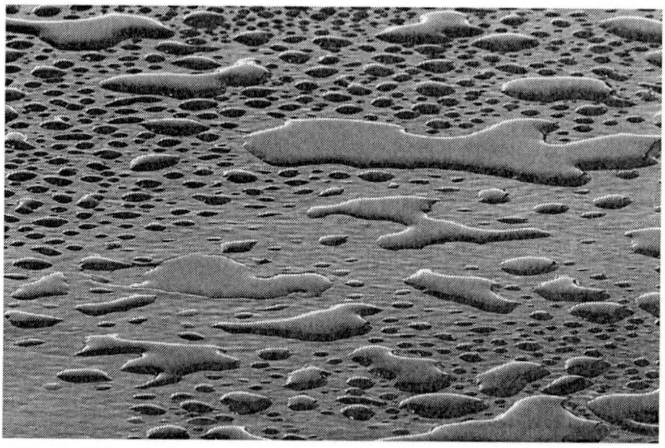

Figure 8: Small puddles of water on a smooth clean surface have perceptible thickness.

The same surface tension demonstration can be done with water, lime water or even saline, but only on a surface made of a substance that the water does not adhere to. Wax is such a substance. Water poured onto a smooth, flat, horizontal wax surface, say a waxed sheet of glass, will behave similarly to the mercury poured onto glass.

The thickness of a puddle of liquid on a surface whose contact angle is 180° is given by:

$$h = 2\sqrt{\frac{\gamma}{g\rho}}$$

where

h is the depth of the puddle in centimetres or metres.

γ is the surface tension of the liquid in dynes per centimetre or newtons per metre.

g is the acceleration due to gravity and is equal to 980 cm/s^2 or 9.8 m/s^2

ρ is the density of the liquid in grams per cubic centimetre or kilograms per cubic metre

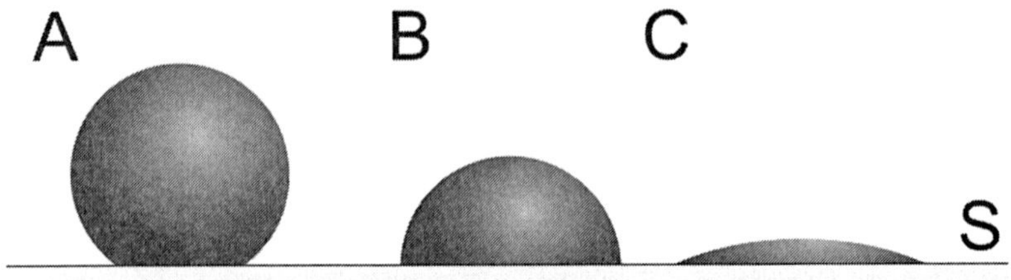

Figure 9: Illustration of how lower contact angle leads to reduction of puddle depth

In reality, the thicknesses of the puddles will be slightly less than what is predicted by the above formula because very few surfaces have a contact angle of 180° with any liquid. When the contact angle is less than 180°, the thickness is given by:

$$h = \sqrt{\frac{2\gamma_{la}(1-\cos\theta)}{g\rho}}$$

For mercury on glass, γ_{Hg} = 487 dyn/cm, ρ_{Hg} = 13.5 g/cm^3 and θ = 140°, which gives h_{Hg} = 0.36 cm. For water on paraffin at 25 °C, γ = 72 dyn/cm, ρ = 1.0 g/cm^3, and θ = 107° which gives h_{H2O} = 0.44 cm.

The formula also predicts that when the contact angle is 0°, the liquid will spread out into a micro-thin layer over the surface. Such a surface is said to be fully wettable by the liquid.

The Breakup of Streams Into Drops

In day-to-day life we all observe that a stream of water emerging from a faucet will break up into droplets, no matter how smoothly the stream is emitted from the faucet. This is due to a phenomenon called the Plateau–Rayleigh instability, which is entirely a consequence of the effects of surface tension.

The explanation of this instability begins with the existence of tiny perturbations in the stream. These are always present, no matter how smooth the stream is. If the perturbations are resolved into sinusoidal components, we find that some components grow with time while others decay with time. Among those that grow with time, some grow at faster rates than others. Whether a component decays or grows, and how fast it grows is entirely a function of its wave number (a measure of how many peaks and troughs per centimetre) and the radii of the original cylindrical stream.

Thermodynamics

As stated above, the mechanical work needed to increase a surface is $dW = \gamma dA$. Hence at constant temperature and pressure, surface tension equals Gibbs free energy per surface area:

$$\gamma = \left(\frac{\partial G}{\partial A}\right)_{T,P,n}$$

where *G* is Gibbs free energy and *A* is the area.

Thermodynamics requires that all spontaneous changes of state are accompanied by a decrease in Gibbs free energy.

From this it is easy to understand why decreasing the surface area of a mass of liquid is always spontaneous ($\Delta G < 0$), provided it is not coupled to any other energy changes. It follows that in order to increase surface area, a certain amount of energy must be added.

Gibbs free energy is defined by the equation, ,
where *H* is enthalpy and *S* isentropy. Based upon this and the fact that surface tension is Gibbs free energy per unit area, it is possible to obtain the following expression for entropy per unit area:

$$\left(\frac{\partial \gamma}{\partial T}\right)_{A,P} = -S^A$$

Kelvin's Equation for surfaces arises by rearranging the previous equations. It states that surface enthalpy or surface energy (different from surface free energy) depends both on surface tension and its derivative with temperature at constant pressure by the relationship.

$$H^A = \gamma - T\left(\frac{\partial \gamma}{\partial T}\right)_P$$

Thermodynamics of Soap Bubble

The pressure inside an ideal (one surface) soap bubble can be derived from thermodynamic free energy considerations. At constant temperature and particle number, $dT = dN = 0$, the differential Helmholtz free energy is given by

$$dF = -PdV + \gamma dA$$

where P is the difference in pressure inside and outside of the bubble, and γ is the surface tension. In equilibrium, $dF = 0$, and so,

$$PdV. = \gamma dA$$

For a spherical bubble, the volume and surface area are given simply by

$$V = \frac{4}{3}\pi R^3 \rightarrow dV \approx 4\pi R^2 dR,$$

and

$$A = 4\pi R^2 \rightarrow dA \approx 8\pi R dR \cdot$$

Substituting these relations into the previous expression, we find

$$P = \frac{2}{R}\gamma,$$

which is equivalent to the Young-Laplace equation when $R_x = R_y$. For real soap bubbles, the pressure is doubled due to the presence of two interfaces, one inside and one outside.

Influence of Temperature

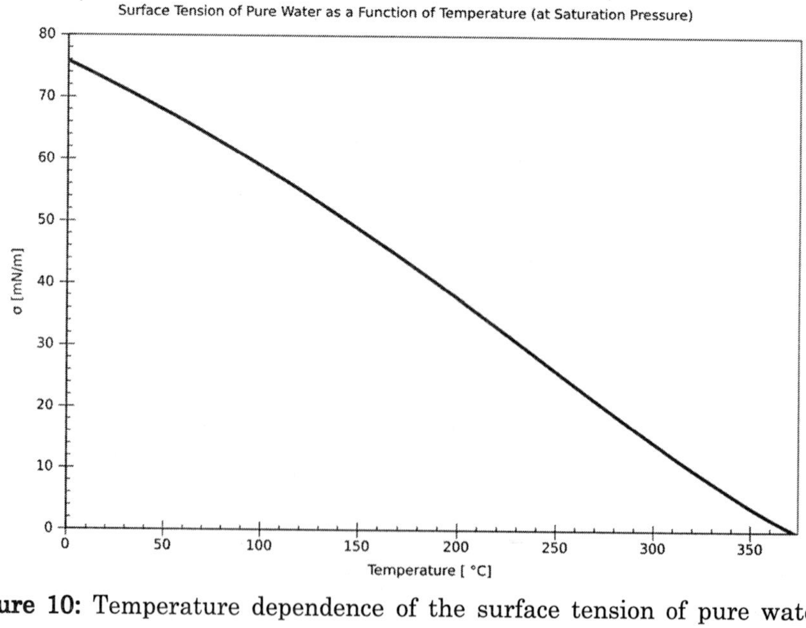

Figure 10: Temperature dependence of the surface tension of pure water

Surface tension is dependent on temperature. For that reason, when a value is given for the surface tension of an interface, temperature must be explicitly stated. The general trend is that surface tension decreases with the increase of temperature, reaching a value of 0 at the critical temperature. There are only empirical equations to relate surface tension and temperature:

- Eötvös:

$$\gamma V^{2/3} = k(T_C - T)$$

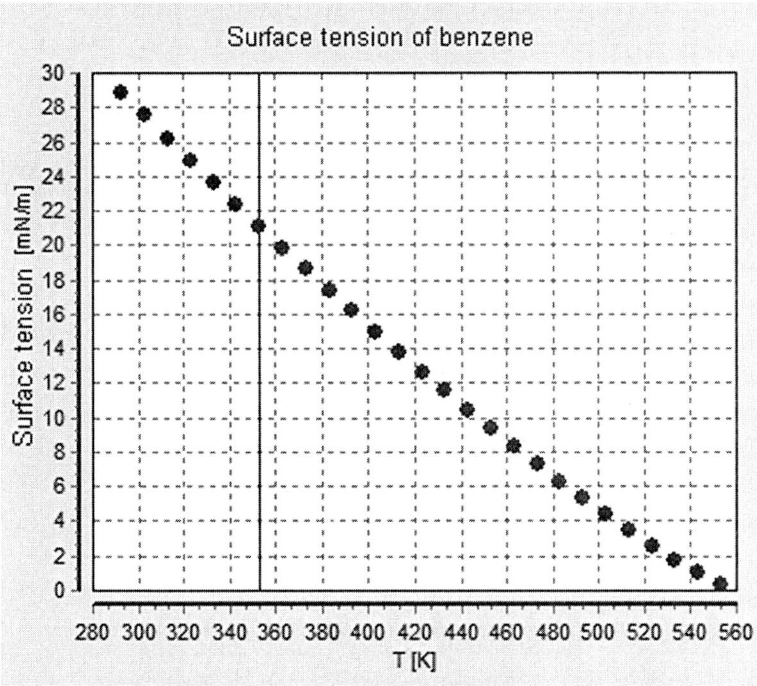

Figure 11: Temperature dependency of the surface tension of benzene

Here V is the molar volume of a substance, T_C *is the critical temperature and* k *is a constant valid for almost all substances. A typical value is* k = *2.1 x* 10^{-7} *[J* $K^{-1}mol^{-2/3}$*]. For water one can further use* V = *18 ml/mol and* T_C = *374°C.*

A variant on Eötvös is described by Ramay and Shields:

$$\gamma V^{2/3} = k\left(T_C - T - 6\right)$$

where the temperature offset of 6 kelvins provides the formula with a better fit to reality at lower temperatures.

- Guggenheim-Katayama:

$$\gamma = \gamma^{o}\left(1 - \frac{T}{T_C}\right)^{n}$$

γ^{o} is a constant for each liquid and n is an empirical factor, whose value is 11/9 for organic liquids.

This equation was also proposed by van der Waals, who further proposed that γ^{o} could be given by the expression, $K_2 T_c^{\frac{1}{3}} P_c^{\frac{2}{3}}$, where K_2 is a universal constant for all liquids, and P_c is the critical pressure of the liquid (although later experiments found K_2 to vary to some degree from one liquid to another).

Both Guggenheim-Katayama and Eötvös take into account the fact that surface tension reaches 0 at the critical temperature, whereas Ramay and Shields fails to match reality at this endpoint.

Influence of Solute Concentration

Solutes can have different effects on surface tension depending on their structure:

- Little or no effect, for example sugar
- Increase surface tension, inorganic salts
- Decrease surface tension progressively, alcohols
- Decrease surface tension and, once a minimum is reached, no more effect: surfactants

What complicates the effect is that a solute can exist in a different concentration at the surface of a solvent than in its bulk. This difference varies from one solute/solvent combination to another.

Gibbs isotherm states that:

$$\Gamma = -\frac{1}{RT}\left(\frac{\partial \gamma}{\partial \ln C}\right)_{T,P}$$

- Γ is known as surface concentration, it represents excess of solute per unit area of the surface over what would be present if the bulk concentration prevailed all the way to the surface. It has units of mol/m^2
- C is the concentration of the substance in the bulk solution.
- R is the gas constant and T the temperature

Certain assumptions are taken in its deduction, therefore Gibbs isotherm can only be applied to ideal (very dilute) solutions with two components.

Influence of Particle size on Vapour Pressure

The Clausius-Clapeyron relation leads to another equation also attributed to Kelvin. It explains why, because of surface tension, the vapour pressure for small droplets of liquid in suspension is greater than standard vapour pressure of that same liquid when the interface is flat. That is to say that when a liquid is forming small droplets, the equilibrium concentration of its vapour in its surroundings is greater. This arises because the pressure inside the droplet is greater than outside.

$$P_v^{fog} = P_v^o e^{\frac{V2\gamma}{RTr_k}}$$

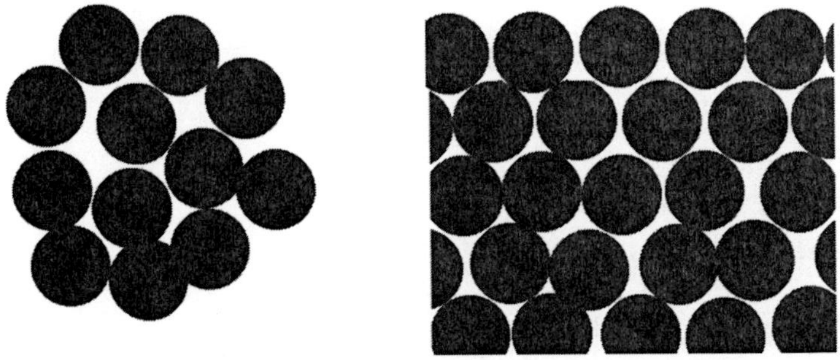

Figure 13: Molecules on the surface of a tiny droplet (left) have, on average, fewer neighbours than those on a flat surface (right). Hence they are bound more weakly to the droplet than are flat-surface molecules.

- P_v^o is the standard vapour pressure for that liquid at that temperature and pressure.
- V is the molar volume.
- R is the gas constant
- r_k is the Kelvin radius, the radius of the droplets.

The effect explains supersaturation of vapours. In the absence of nucleation sites, tiny droplets must form before they can evolve into larger droplets. This requires a vapour pressure many times the vapour pressure at the phase transition point.

This equation is also used in catalyst chemistry to assess mesoporosity for solids.

The effect can be viewed in terms of the average number of molecular neighbours of surface molecules.

The below shows some calculated values of this effect for water at different drop sizes:

Table 1: P/P_0 for water drops of different radii at STP

Droplet radius (nm)	1000	100	10	1
P/P_0	1.001	1.011	1.114	2.95

The effect becomes clear for very small drop sizes, as a drop of 1 nm radius has about 100 molecules inside, which is a quantity small enough to require a quantum mechanics analysis.

Data Table

Table 2: Surface tension of various liquids in dyn/cm against air Mixture %'s are by massdyn/cm is equivalent to the SI units of mN/m (milli-Newton per metre)

Liquid	*Temperature °C*	*Surface Tension, γ*
Acetic acid	20	27.6
Acetic acid (40.1%) + Water	30	40.68
Acetic acid (10.0%) + Water	30	54.56
Acetone	20	23.7
Diethyl ether	20	17.0
Ethanol	20	22.27
Ethanol (40%) + Water	25	29.63
Ethanol (11.1%) + Water	25	46.03
Glycerol	20	63
n-Hexane	20	18.4
Hydrochloric acid 17.7M aqueous solution	20	65.95
Isopropanol	20	21.7
Mercury	15	487
Methanol	20	22.6
n-Octane	20	21.8
Sodium chloride 6.0M aqueous solution	20	82.55
Sucrose (55%) + water	20	76.45
Water	0	75.64
Water	25	71.97
Water	50	67.91
Water	100	58.85

Bibliography

Alfven, H.: *Cosmical Electrodynamics*, Oxford Univ. Press, Oxford, 1950.

Antsiferov, V. V. and G. I. Smirnov: *Coherent Radiation Processes in Plasmas*, Cambridge Univ. Press, Cambridge, 1999.

Birdsall, C. K. and A. B. Langdon: *Plasma Physics via Computer Simulation*, McGraw-Hill, NY, 1991.

Braams, C. and P. Stott: *Nuclear Fusion: Half a Century of Magnetic Confinement Fusion*, IOP Press, NJ, 2002.

Brambilla, M. Kinetic: *Theory of Plasma Waves: Homogeneous Plasmas*, Oxford Press, Oxford, 1998.

Buchner, J.: *Physics of Space Plasmas*, MIT Press, US, 1995.

Charles F. Kennel, *Convection and Substorms*, Oxford Univ. Press, Oxford, 1995.

Chen, Francis F.: *Plasma Physics and Controlled Fusion*, Plenum Press, London, 1984.

Choudhuri, A.: *The Physics of Fluids and Plasmas: An Intro for Astrophysicists*, Cambridge Univ. Press, Cambridge, 1999.

Cross, R. C.: *An Introduction to Alfven Waves*, Hilger, Bristol, 1988.

Daniel Hastings, Henry Garrett: *Spacecraft-Environment Interactions*, Cambridge University Press, Cambridge, 2000.

Das, A. C.: *Space Plasma Physics: An Introduction,* Narosa Publishing House, New Delhi, 2004.

Davidson, R.: *The Physics of Nonneutral Plasmas*, Imperial College Press, London, 2001.

Davidson, Ronald: *An Introduction to the Physics of Nonneutral Plasma*, Addison-Wesley, London, 1990.

Dendy, R. O.: *Plasma Dynamics*, Oxford University Press, Oxford, 1990.

Dwight Nicholson: *Introduction to Plasma Theory*, Wiley, NJ, 1983.

Elskens, Y. and D. Escande: *Microscopic Dynamics of Plasmas and Chaos*, IOP, NJ, 2002.

Francis Chen: *Introduction to Plasma Physics and Controlled Fusion,* Plenum Press, London, 1974.

Freidberg, J. P.: *Ideal Magnetohydrodynamics*, Plenum Pr., NY, 1987.

Freidberg, Jeffrey: *Plasma Physics and Fusion Energy*, Cambridge Univ. Press, Cambridge, 2007.

Fujimoto, Takashi: *Plasma Spectroscopy,* Clarendon Press, Oxford, 2004.

Galeev, A. and R. Sudan: *Basic Plasma Physics*, North-Holland, 1989

Gareth Wynn-Williams: *The Fullness of Space,* Cambridge Press, Cambridge, 1992.

Gerd W. Prolss: *Physics of the Earth's Space Environment,* Springer-Verlag, London, 2004.

Gibbon, Paul: *Short Pulse Laser Interactions with Matter*, Imperial College Press, London, 2000.

Goedbloed, J. and S. Poedts: *Principles of Magnetohydrodynamics: With Applications to Laboratory and Astrophysical Plasmas,* Cambridge Press, Cambridge, 2004.

Goldston, R. J. and P. H. Rutherford: *Introduction to Plasma Physics,* IOP, NJ, 1995.

Gorse, C.: *Plasma Technology: Fundamentals and Applications*, Plenum Press, NY, 1992.

Gunther Rudiger and Rainer Hollerbach: *The Magnetic Universe: Geophysical and Astrophysical Dynamo Theory* , Wiley-VCH, 2004.

Gurnett, D. A. and A. Bhattacharjee: *Introduction to Plasma Physics with Space and Laboratory Applications,* Cambridge Univ. Press, Cambridge, 2005.

Hans Wilhelmsson: *Fusion: A Voyage Through the Plasma Universe,* IOP, 1999.

Hans, R.Griem: *Principles of Plasma Spectroscopy*,Cambridge Press, Cambridge, 1997.

Harms, A. A.: *Principles of Fusion Energy,* World Scientific, NJ, 2000.

Hazeltine, R. and F. Waelbroeck: *The Framework of Plasma Physics,* Perseus Books, NY, 1998.

Hazeltine, R.D.; Waelbroeck, F.L.: *The Framework of Plasma Physics*, Westview Press, US, 2004.

Hitchon, W. N. G.: *Plasma Processes for Semiconductor Fabrication*, Cambridge Univ. Press, Cambridge, 1999.

Horton, W. and Y-H Ichikawa: *Chaos and Structures in Nonlinear Plasmas*, World Scientific, 1996.

John W. Freeman: *Storms in Space,* Cambridge, 2001.

Kallenrode, M.B.: *Space Physics: Plasmas and particles in the Heliosphere and Magnetosheres*, Springer, London, 1998.

Kenneth R. Lang: *Sun, Earth and Sky,* Springer-Verlag, Berlin, 1995, 1997.

Kenneth, T. Fowler: *The Fusion Quest,* Johns Hopkins Press, Baltimore, MD, 1997.

Kenro Miyamoto, Plasma Physics for Nuclear Fusion, MIT Press, US, 1980, 1989.

Kirk, J. G.; D. B. Melrose, and E. R. Priest: *Plasma Astrophysics*, Springer-Verlag, London, 1994.

Kivelson, Margaret and Chris Russell: *Introduction to Space Physics*, Cambridge Univ. Press, Cambridge, 1995.

Krall, N. and A. Trivelpiece: *Principles of Plasma Physics*, McGraw-Hill, NY, 1973.

Leonard F. Burlaga: *Interplanetary Magnetohydrodynamics*, Oxford Univ. Press, Oxford, 1995.

Lequeux, James: *The Interstellar Medium,* Springer-Verlag, London, 2005.

Lindl, J.: *Inertial Confinement Fusion*, Springer, NJ, 1998.

Liu, C. S. and V. K. Tripathi: *Interaction of Electromagnetic Waves with Electron Beams and Plasmas*, World Scientific, NJ, 1995

Lochte-Holtgreven, W.: *Plasma Diagnostics*, North-Holland, 1968.

Manheimer, W. and C. Lashmore-Davies: *MHD and Microinstabilities in Confined Plasmas*, IOP, NJ, 1989.

McCracken, G. and P. Stott: *Fusion: The Energy of the Universe*, Elsevier, London, 2005.

Melrose, Donald: *Instabilities in Space and Laboratory Plasmas*, Cambridge Univ. Press, Cambridge, 1986.

Mendonca, J. T.: *Theory of Photon Acceleration*, IOP, NJ, 2000.

Mikahilovskii, A. B.: *Instabilities in a Confined Plasma*, IOP, NJ, 1998.

Moiseev, S. S.; V. Oraevsky, and V. Pungin: *Non-Linear Instabilities in Plasmas and Hydrodynamics*, IOP Press, NJ, 1999.

Nishikawa, K. and M. Wakatani: *Plasma Physics: Basic Theory with Fusion Applications*, Springer-Verlag, London, 1990.

Oks, E. A.: *Plasma Spectroscopy*, Springer-Verlag, London, 1995.

Paul D. Thompson: *Gases & Plasmas,* Lippincott Company, Philadelphia, 1966.

Paul, M. Bellan, Spheromaks, Imperial College Press, London, 2000.

Per Helander and Dieter J. Sigmar: *Collisional Transport in Magnetized Plasmas*, Cambridge Press, Cambridge, 2002.

Peratt, Anthony: *Physics of the Plasma Universe*, Springer-Verlag, London, 1992.

Plasma Chemistry, L. S. Polak and Yu A. Lebedev, eds., Cambridge, 1999.

Richard Dendy: *Plasma Physics,* Cambridge Press, Cambridge, 1993.

Rozhansky, V. A. and L. D. Tsendin: *Transport Phenomena in Partially Ionized Plasma*, Taylor & Francis Group, London, 2001.

Sanborn C. Brown: *Basic data of Plasma Physics*, AIP Press, NY, 1994.

Sazhin, Sergei: *Whistler-mode Waves in a Hot Plasma*, Cambridge Univ. Press, Cambridge, 1993.

Shevelko, V. and L. Vainshtein: *Atomic Physics for Hot Plasmas*, Oxford Press, Oxford, 1993.

Stangeby, P. C.: *The Plasma Boundary of Magnetic Fusion Devices*, IOP Press, NJ, 2000.

Steven T. Suess and Bruce T. Tsurutani: *From the Sun: Auroras, Magnetic Storms, Solar Flares, Cosmic Rays,* American Geophysical Union, 1998.

Sturrock, P.: *Plasma Physics: An Intro to the Theory of Astrophysical, Geophysical, and Laboratory Plasmas*, Cambridge Univ. Press, Cambridge, 2004.

Tajima, T. and K. Shibata: *Plasma Astrophysics,* Addison-Wesley, NY, 1997.

Tajima, T.: *Computational Plasma Physics with Applications to Fusion and Astrophysics*, Westview Press, US, 2004.

Thomas Stix: Waves in Plasmas, AIP Press, NY, 1992.

Toshiro Ohnuma: *Radiation Phenomena in Plasmas*, World Scientific, NJ, 1996.

van Roosmalen, A. J.; J. A. G. Baggerman and S.J.H. Brader: *Dry Etching for VLSI,* Plenum Press, NY, 1991.

Verheest, F.: *Waves in Dusty Space Plasmas*, Kluwer, UK, 2000.

Vladimir, Fortov and Igor Iakubov: *The Physics of Non-Ideal Plasma,* Imperial College Press, London, 1999.

Wakatani, M.: *Stellerator and Heliotron Devices,* Oxford Univ. Press, Oxford, 1998.

Wesson, J.: *Tokomaks,* Oxford Univ. Press, Oxford, 2004.

Wielebinski, R. and R. Beck: *Cosmic Magnetic Fields*, Springer-Verlag, London, 2005.

Wilhelmsson, H. and E. Lazzaro: *Reaction-Diffusion in the Physics of Hot Plasmas*, IOP, NJ, 2000.

William Kruer: *The Physics of Laser Plasma Interactions*, Westview Press, US, 2003.

Wolfgang Kundt: *Astrophysics: A New Approach,* Springer-Verlag, London, 2005.

Yaffa & Shalom Eliezer: *The Fourth State of Matter,* Hilger, Bristol, 1989.

Index

G

H

I

K

L

M

N

O

P

R

S

T

U

V

W

X

❑❑❑